Topical Conference: Waste Plastics

Held at the 2023 AIChE Annual Meeting

Orlando, Florida, USA
5-10 November 2023

ISBN: 978-1-7138-9312-7

Topical Conference: Waste Plastics

Held at the 2023 AIChE Annual Meeting

Orlando, Florida, USA
5-10 November 2023

AIChE
The Global Home of Chemical Engineers

Printed from e-media with permission by:

Curran Associates, Inc.
57 Morehouse Lane
Red Hook, NY 12571

CURRAN ASSOCIATES INC.
proceedings
.com

Some format issues inherent in the e-media version may also appear in this print version.

AIChE
120 Wall Street, FL 23
New York, NY 10005-4020

Phone: (800) 242-4363
Fax: (203) 775-5177

www.aiche.org

Additional copies of this publication are available from:

Curran Associates, Inc.
57 Morehouse Lane
Red Hook, NY 12571 USA
Phone: 845-758-0400
Fax: 845-758-2633
Email: curran@proceedings.com
Web: www.proceedings.com

TABLE OF CONTENTS

ANALYSIS AND ASSESSMENT FOR SOLVING THE PLASTIC WASTE CRISIS

CONVERSION OF WASTE PLASTIC INTO LIQUID FUELS

PLENARY SESSION: WASTE PLASTICS (INVITED TALKS)

POSTER SESSION: WASTE PLASTICS

RECYCLING AND UPCYCLING OF PLASTIC WASTE

239a Waste-to-Energy and Chemical Recycling of Mixed Plastic Waste. Economic and Environmental Optimisation of the Northern Italian Supply Chain

Daniel Cristiu, Federico d'Amore, Fabrizio Bezzo

1. Problem

Plastics are the world's largest synthetic consumer product, and their increasing demand poses a challenge in terms of end-of-life management (recycling, energy recovery, landfilling), an issue that needs to be addressed in the transition to a circular economy. The same resilient properties that lead to high annual production, reaching 390 million tonnes in 2021 (Plastic Europe, 2022), also create a major environmental problem in terms of end-of-life management, as significant fractions of plastic waste that are not properly disposed of most often end up in landfills or leak into the ocean, with harmful consequences (Li et al., 2022). The largest fraction of plastic waste is made up of plastic packaging; only a limited fraction can be mechanically recycled due to downcycling, and the residual fraction, called mixed plastic waste (MPW), which is often landfilled, can be turned into a resource according to a circular economy approach. The challenges in managing MPW are due to variability in its composition, and incineration can help reduce the amount of waste that needs to be landfilled and generate electricity. Pyrolysis and gasification represent potentially attractive emerging technologies for MPW management (Solis and Silveira, 2020). In this study, the waste-to-energy pathway is represented by incineration with electricity generation, and the chemical recycling pathways are represented by gasification, producing syngas from which electricity is generated, and pyrolysis, generating a pyrolysis oil to be transported to a refinery for further processing. In this context, it is important to explore the potential of waste-to-energy technologies and chemical recycling for the management of mixed plastic waste and to assess them based on the economic feasibility and environmental impact. Supply Chain (SC) optimisation can be used to evaluate both the optimal selection of MPW treatment technologies and the optimal distribution of material flows between the different stages of the SC. This study aims to propose a decision-making tool for MPW supply chain management through a multi-objective, spatially explicit Mixed integer linear programming (MILP) framework. The model aims at finding the trade-offs between economic optimisation (in terms of maximising the annual gross profit) and environmental impact (in terms of minimising the net GHG emissions of the supply chain), focussing on Northern Italy.

2. Methodology and expected results

The problem is formulated as an annualised spatially-explicit MILP modelling framework. The spatially-explicit nodes are described in the model through a set n comprising 47 provinces, 12 sorting centres, 24 incinerators and 3 refineries. Each node is characterised by its exact geographic coordinates. Gasification plants can be installed only in proximity of incineration plants; similarly, pyrolysis plants can be placed either close to incineration plants or refineries. Both waste-to-energy (i.e. incineration) and chemical recycling (i.e. pyrolysis and gasification) technologies are considered as treatment options. The stages of the MPW SC are divided into the transport of plastic packaging waste (PPW) from provinces to sorting centres, where the PPW share that can be mechanically recycled is separated from the residual part, i.e., the MPW. From the sorting centres, the MPW fraction is transported to a treatment plant (incineration, gasification, pyrolysis), whose outputs are: electric power in the case of incineration and gasification, and oil in the case of pyrolysis plants, the latter one to be transported to a refinery for further processing. Road transport is considered as the only transport option since it reflects current practice for plastic waste in Northern Italy.

The modelling framework aims at maximising the economic performance of the overall SC from the perspective of the final treatment plants (in terms of gross profit) and minimising the environmental impact of the supply chain (in terms of net greenhouse gas emissions).

The decision variables in the optimization problem are:

- the PPW flowrates distribution from the provinces to the sorting centres;
- the MPW flowrates distribution from the sorting centres to the treatment plants;
- the selection of technology, plant size and location of the treatment plants;
- the pyrolysis oil flowrate distribution from a pyrolysis plant to a refinery.

The modelling framework provides a decision tool for the optimal choice of the treatment technology for MPW in terms of both economic performance and environmental impact and can provide different trade-offs between the conflicting objectives.

3. Conclusions

This study proposes a multi-objective mixed integer linear programming framework for the Northern Italian supply chain of mixed plastic waste management. The model optimisation outcomes provide quantitative information on the SC in terms of economic and environmental impact performance, and provide the optimal selection of the treatment technology under different scenarios. Results show that, when aimed for economic optimisation, the selected MPW treatment technology is exclusively the incineration, being the technology that has the highest gross profit, while in the case of environmental optimisation, the selected technology is pyrolysis, due to lower GHG emissions of the technology.

References

1. Li, et al., 2022, Expanding Plastics Recycling Technologies: Chemical Aspects, Technology Status and Challenges, Green Chemistry Journal, 1463-9262

Plastics Europe (2022). Plastics – The facts 2022. [Accessed 29/03/2023] (Available at: https://plasticseurope.org/knowledge-hub/plastics-the-facts-2022).

105. Solis, S. Silveira, 2020, Technologies for chemical recycling of household plastics – A technical review and TRL assessment, Waste Management, 105.

239b Comparative Techno-Economic and Lifecycle Assessment of Sustainably Repairable r-PET Plastics.

Vedanti Sharma, Mark Mba Wright, Dhananjay Dileep, Eric Cochran

PET is a commonly used plastic for packaging, particularly in the bottling and beverage industries. Since these petroleum-based plastics have a long life span and do not decompose naturally, they contribute to the accumulation of 'white pollution' when not disposed of properly. Due to this and inadequate waste disposal methods, they can leak into the environment. The production of PET was approximately 81 million metric tons in 2021 and is predicted to increase to 115 million tons by 2028 due to the rising global population and subsequent increase in food consumption. Despite extensive efforts to recycle bottles, only a small fraction of PET is actually recycled, and even recycled PET is frequently discarded in landfills. Furthermore, only 7% is utilized for bottle-to-bottle recycling. The experimental data used in the analysis was obtained by introducing an intelligently engineered moiety nicknamed Trojan Horse units that specifically cleave under milder conditions than traditional PET. The reactive centers initiate the PET chain's demolition, which is expected to progress till re-polymerizable oligomers or monomers.

This research conducted a techno-economic analysis (TEA) and lifecycle assessment (LCA) of REPAIR-Enabled Environmental Degradation agents to evaluate their potential in creating economically viable and sustainable r-PET plastics. The study utilized a chemical process model based on the National Renewable Energy Laboratory (NREL) biorefinery design and BioSTEAM to benchmark the analysis against recent NREL studies.

The analysis yielded approximately $1.46/kg for the cost of traditional PET, along with 4.74 kg CO2eq/kg for emissions. However, recovering half of the discarded PET and introducing 25% BioTPA and 5% TH reduces emissions to 2.6 kg CO2eq/kg, accompanied by an increased cost of $1.84/kg. On the other hand, if the recovery rate is raised to 90%, the cost drops to $1.18/kg, while emissions decrease to 0.52 kg CO2eq/kg. These findings imply that there might be a threshold point where r-PET becomes both more affordable and sustainable than fossil-based plastics.

239c From Cradle to Grave: A Comprehensive Study of the Carbon Emissions of Aircraft Cabin Interiors through Life Cycle Assessment

Su Natasha Binti Mohamad, Rachael Rothman

A large amount of polymers and plastic composite are used in the manufacturing and production of nowadays modern aircrafts in an effort to achieve the desired light weighting airframe whilst sustaining robust performance. Plastics present an attractive option for airframe construction due to their exceptional properties, notably their high resistance to corrosion and fatigue. This resistance to corrosion is especially advantageous in harsh environmental conditions, making plastics an optimal choice for aircraft that are required to operate in such conditions. Moreover, the ability to engineer plastics with good fatigue resistance enables them to withstand repeated loading cycles without the risk of crack propagation or failure. Therefore, utilising plastics alternative in building the aircraft frames and interiors presents a practical solution for reducing fuel consumption and decreasing production costs when compared to traditional metal-based airframe. Unfortunately, due to wide range of plastic materials used in airframe construction, this renders their segregation and recycling problematic once the aircraft reaches the end of its useful life and is no longer deemed safe or financially viable to continue operating. As a consequence, this create a heterogeneous mixture of plastic waste where landfill disposal and energy recovery have become a prevalent waste management practices for these plastic materials due to the difficulties associated with accurately identifying and segregating the specific types of plastics. The intricate composition of these plastic amalgams presents significant hurdles for the attainment of sustainable exit for these polymers and further, the realization of a circular economy.

Therefore, in this study, assorted plastic parts from the cabin interior of an aircraft fuselage including passenger safety unit, window panes, in-flight entertainment cover and etc. were meticulously characterised according to dimensions, colour, weight and polymer composition using Fourier-Transform Infrared Spectroscopy (FTIR). FTIR results identified polyetherimide, polycarbonate, nylon, polyvinylchloride, silicon rubber, polymethyl-methacrylate and polysulfone among other types of plastics. This finding served as the primary data in a life cycle assessment (LCA) inventory analysis where LCA was carried out to assess the environmental burden and sustainability performance of the plastics used in the aircraft cabin interior from cradle to grave using SimaPro 9.4.0.2 software. Secondary data is supported by Ecoinvent 3 database and literature. This LCA was conducted in accordance to ISO standard 14040 and ISO 14044, where the goal and scope of the study (i.e., system boundary, functional unit, impact categories, assessment method, data quality and etc.) were properly defined. Our preliminary finding postulated a hypothesis where the carbon footprint of the embodied carbon of plastics are considerably greater than the greenhouse gas (GHG) emission of fuel consumption. The discovery enabled us to identify the environmental hotspots and cost-benefit trade-offs of material selection in the pursuit of creating an environmentally sustainable airframe.

239d Environmental and Economic Analyses of Chemical Recycling Via Dissolution of Polyethylene Terephthalate

Utkarsh Chaudhari, Daniel Kulas, Alejandra Peralta, Tasmin Hossain, Anne Johnson, Damon Hartley, Robert Handler, Barbara Reck, Vicki Thompson, David W. Watkins, David Shonnard

Plastics have become an integral part of our day-to-day life owing to their low cost and excellent material properties suitable for multiple applications in several manufacturing industries. High consumption and production of plastics with low collection rates has led to a global plastic waste crisis. Unfortunately, the plastics supply chains globally, and in the U.S., are highly dependent on fossil resources, leading to high greenhouse gas (GHG) emissions and consumption of energy. Polyethylene terephthalate (PET, resin code #1) is one of the important plastic materials and finds applications in production of bottles, fibers (clothes, carpets etc.), films, sheets, and thermoforms. In the U.S., PET waste accounts for ~15% of the total plastic waste produced, most of which is currently being landfilled (66%), incinerated with energy recovery (15%), and the remaining is collected for sorting and recycling (19%). The complete PET supply chain processes in the U.S. released 23 MMT CO_2-eq and consumed 522 PJ of energy in 2019. These emissions represented 1.2% of the total U.S. industry related GHG emissions in 2019[1].

To break this linearity of plastic supply chains, emerging chemical recycling technologies have been proposed to be a part of this solution to implement circular economy by replacing fossil derived plastic resins. These technologies can be broadly classified in 3 main categories: dissolution/purification, depolymerization, and conversion recycling technologies[2]. While the economic and environmental impacts of enzymatic and non-enzymatic depolymerization of waste PET, as well as dissolution of PET using non-green solvents, are known from the literature, these impacts remain unknown for the dissolution process using a green solvent. In addition, none of the previously published studies have evaluated and compared these impacts for PET specific dissolution process with polymer recovery by the evaporation approach. Also, the economic metrics such as net present value, payback period, return on investment, discounted internal rate of return for these processes are not reported in the literature.

Therefore, the purpose of our study was to address these research gaps and compare the environmental and economic performance of dissolution recycling technology for waste PET using a green renewable solvent[3], Gamma (γ)-Valerolactone (GVL), with different polymer recovery techniques. Based on our previously published systems analysis framework[2, 4], we take a "bottom-up" approach, in which process simulation (ChemCAD) is integrated with life cycle assessment (LCA) and techno-economic analysis (TEA). The **first research objective** of our study was to conduct process simulation of chemical recycling of waste PET via dissolution process with the solvent GVL. We analyze three different processes that differ in their polymer precipitation techniques, as mentioned above, using the same solvent. The **second research objective** was to evaluate technoeconomic performance metrics of the simulated recycling processes. We also look at the effect of scaling up the capacity on the total capital costs and minimum selling price. We also couple this economy of scale analysis with a simple transportation model to account for transportation costs over longer distances with increasing capacity, which has been neglected in prior TEA studies for chemical recycling technologies. The **third research objective** was to evaluate environmental metrics such as GHG emissions and cumulative energy demand (CED) of simulated recycling processes.

Our analysis revealed that the choice of polymer precipitation method affects the overall economic and environmental performance of the process. The MSP of CR-PET, at 8,400 MT/year, was found to range from $1.269 to $1.532/kg of CR-PET, which was lower than fossil derived PET. The "cradle-to-gate" GHG emissions for PET dissolution processes ranged from 1.09-3.56 kg CO_2-eq/kg of CR-PET, with that being the highest for anti-solvent process. Similarly, the total energy consumption was found to range from 17.95 – 56.31 MJ/kg of CR-PET. Overall, dissolution of PET with polymer precipitation via evaporation and cooling methods were found to be economically and environmentally favorable than anti-solvent method as well as fossil derived PET resin.

References

1. Chaudhari, U.S., Johnson, A.T., Reck, B.K., Handler, R.M., Thompson, V.S., Hartley, D.S., Young, W., Watkins, D. and Shonnard, D., 2022. Material Flow Analysis and Life Cycle Assessment of Polyethylene Terephthalate and Polyolefin Plastics Supply Chains in the United States. *ACS Sustainable Chemistry & Engineering*, *10*(39), pp.13145-13155.

2. Chaudhari, U.S., Lin, Y., Thompson, V.S., Handler, R.M., Pearce, J.M., Caneba, G., Muhuri, P., Watkins, D. and Shonnard, D.R., 2021. Systems analysis approach to polyethylene terephthalate and olefin plastics supply chains in the circular economy: A review of data sets and models. *ACS Sustainable Chemistry & Engineering*, *9*(22), pp.7403-7421.

3. Chen, W., Yang, Y., Lan, X., Zhang, B., Zhang, X. and Mu, T., 2021. Biomass-derived γ-valerolactone: Efficient dissolution and accelerated alkaline hydrolysis of polyethylene terephthalate. *Green Chemistry*, *23*(11), pp.4065-4073.

4. Shonnard, D., Tipaldo, E., Thompson, V., Pearce, J., Caneba, G. and Handler, R., 2019. Systems analysis for PET and olefin polymers in a circular economy. *Procedia CIRP*, *80*, pp.602-606.

239f Techno-Economic and Life Cycle Analysis of Polypropylene Plastic Waste Recycling through Fast Pyrolysis

Amir Akbari, Sam Barker, Mason Aksomitus, Gary Aurand, Rui Shi

Plastic disposal has been generating several environmental problems in the past decades. Currently, only 9% of plastic waste is mechanically recycled in the United States. Chemical recycling breaks down the plastics waste into its constituents, therefore complements mechanical recycling, reducing the environmental impacts and resource depletion. The pyrolysis process can handle hard-to-recycle plastics, while producing a wide range of valuable products including oil, waxes, and aromatics. Therefore, pyrolysis provides opportunities to reduce oil usage, carbon dioxide emissions, and the quantities of waste for disposal. On the other hand, the economic potential and environmental impacts are not well studied for particular feedstocks.

Polypropylene (PP) is one of the most abundantly used plastic types, and it constitutes 16% of the global plastic production. For this study, we focus on the design, synthesis, and simulation of an open-loop recycling process of waste PP through fast pyrolysis, and exam the environmental and economic performances. The reactor was designed to operate at 510 °C. Operating at this temperature had a synergistic effect on the plant by optimizing the production of paraffin waxes while suppressing the formation of liquid petroleum gas (LPG) and aromatics The subsequent produced waxes were further cracked using a fluid catalytic cracking (FCC) reactor in a second fluidized bed reactor. In addition, a total of three distillation columns were utilized to separate product streams and produce 99 % purity gasoline and diesel. Techno-economic analysis and life-cycle assessment were conducted to characterize the economic feasibility and environmental implications of the proposed technology. The initial dried plastic throughput of the plant was set at 2500 kg/hr, while sensitivity analysis on important process parameters such as plant capacity, and volatility of gasoline, diesel, and plastic prices was done to obtain the optimum plant design conditions for both profitability as well as the environment. Results of this research indicate that a fine-tuned fast pyrolysis process integrated with downstream upgrading and separation units presents a potential opportunity to feasibly convert plastic wastes into liquid fuels such as gasoline and diesel to achieve circular economy.

239g Many Happy Returns - Making Plastic Reuse Mainstream

Rachael Rothman

Plastics are essential to our everyday lives, but their waste, particularly when in an unmanaged waste system, causes great environmental issues. The waste hierarchy is very clear that first we should reduce, then reuse, then recycle. Here we focus on reuse - when, where and how do reusable plastics systems make sense? The Many Happy Returns project has focussed on reuse of food packaging. Environmental and behavioural science along with linguistics are considered to understand how to make reuse mainstream. Life Cycle Assessment shows the importance of container weight, energy source for washing, water use in washing and any transport involved in reverse logistics. Break even points for a selection of environmental impacts, including global warming potential are calculated, showing the number of times a container must be used for it to make environmental sense. Containers are also subjected to an intense wear process, then discoloured and scratched containers shown to consumers using a behavioural psychology methodology to assess if people would be prepared to use them. The overlap of these two methodologies demonstrates if people are prepared to use the containers the required number of times. Linguistics highlights the importance of words to ensure engagement with a system, and how to increase uptake.

239h Compostable Plastics - How Do We Assess Them and Where Should We Use Them?

Rachael Rothman

Compostable plastics present an interesting alternative to traditional fossil based non-degradable plastics, however current assessment techniques do not accurately reflect their potential or environmental impacts. First we discuss the use of life cycle assessment to assess compostable plastics, highlighting gaps in data and knowledge the prevent a full picture being assessed. In particular, databases commonly used for life cycle assessment do not include data for end of life for compostable plastics, instead using a generic "biowaste" data set which includes food waste (and therefore different nutrients and carbon content). Some researchers adapt this base data set to include more accurate CO_2 and CH_4 emissions from compostable plastics, giving a better picture of global warming potential, however it is only emissions to aid that are considered. We discuss the other impacts that are important, including those on the soil and suggest guideline for life cycle assessment that should be followed when considering compostable plastics. In particular we highlight the need to consider additionality - what extra food waste will be composted (and the nutrients therefore captured) if compostable plastics are used?

298a Economic and Environmental Analysis of Plastics Pyrolysis after Secondary Sorting of Mixed Plastic Waste

Daniel Kulas, Ali Zolghadr, Utkarsh Chaudhari, David Shonnard

Plastic usage has skyrocketed in recent years as plastics have become an essential material to everyday life. Unfortunately, recycling rates have not matched the increased growth, with the majority of waste plastic (approximately 80%) currently landfilled in the United States. New chemical recycling technologies, i.e., pyrolysis, are needed to increase the recycling rates of waste plastics. However, pyrolysis faces challenges because of bridging the feed into the hot reactor and with the high cost of waste plastic feedstock. In this work, a novel liquid-fed pyrolysis process, designed to eliminate bridging, was modeled through process simulation using Aspen Plus and then assessed over a range of processing conditions. A secondary sortation of inexpensive mixed plastic waste from material recovery facilities provided a low-cost feedstock (PE, PP) for pyrolysis.

A techno-economic analysis was conducted using a discounted cash flow analysis to analyze the minimal selling price (MSP) of pyrolysis oil as an alternative to fossil naphtha. Naphtha is a commodity chemical feedstock for plastics production. A life cycle assessment was conducted to determine the 'gate to gate' greenhouse gas emissions and energy consumption of pyrolysis oil compared to fossil naphtha using SimaPro. Our analysis found favorable economic and environmental results for pyrolysis oil when compared to fossil naphtha ($561/t, 0.52 kg CO_2 eq./kg). A sensitivity analysis found the economic results most responsive to waste plastics feedstock costs, pyrolysis gas sales, operating capacity, and waste disposal costs. Finally, a comprehensive interrogation of the TEA and LCA parameter space will be presented to provide an understanding into the relationships and trade-offs between the economic and environmental impacts of this novel process. The lowest minimum selling price and greenhouse gas emissions for pyrolysis oil were achieved by recovering and selling sorted PET from the mixed plastic bale and implementing process heat integration surrounding the pyrolysis reactor.

298b Precise Control of Microwave Assisted Pyrolysis for Waste Plastics Via Fluorescent Nano-Thermometers

Zhenyu Zhao, Xin Gao

In the present study, a microwave assisted catalytic pyrolysis technique was investigated for converting waste high-density polyethylene into liquid fuels. During the microwave-assisted catalytic pyrolysis of waste high-density polyethylene, the selective heating characteristic of microwave irradiation causes the formation of local high-temperature domains ("hot spots") over the surface of microwave-absorbing catalysts, leading to rapid cracking of polymers near the catalyst particles and hence the generation of hydrocarbon products. This mechanism has been confirmed by investigating the dominant role of the catalysts' dielectric loss on the composition of produced hydrocarbon. In order to precisely control this "hot spots" phenomenon, we designed thermo-sensitive fluorescent nanoparticles as nano thermometers to probe the intrinsic temperature of iron-based catalyst particles and accordingly derived an energy transfer model to predict the temperature gradient between catalysts and the bulk reactants. The fluorescence results agree with the model predictions that the microwave induced temperature gradient can be enlarged by increasing microwave intensity, as well as the dielectric loss and size of particles. Conversely, the increase in the thermal conductivity and velocity of sweeping gas lowers the temperature gradient. Based on the model, we precisely designed various microwave heating strategies and compared the composition of produced hydrocarbons in these cases. A narrower hydrocarbon product distribution was obtained by using higher-power short-pulse microwave heating strategy to enlarge the catalyst-reactant temperature gradient, exhibiting the unique advantage of microwave irradiation that the instantaneous on-off switch is able to reduce the thermal inertia of processes and the feasibility of the directional conversion of waste plastics to certain high-valued fuel products. Besides, the average chain length of produced hydrocarbons can be flexibly regulated from C_9 to C_{30} by adjusting the microwave power and the sweeping nitrogen velocity. Results from the present work can serve as the starting point for fine process control and rational development of microwave responsive catalysts for catalytic recycling of plastics.

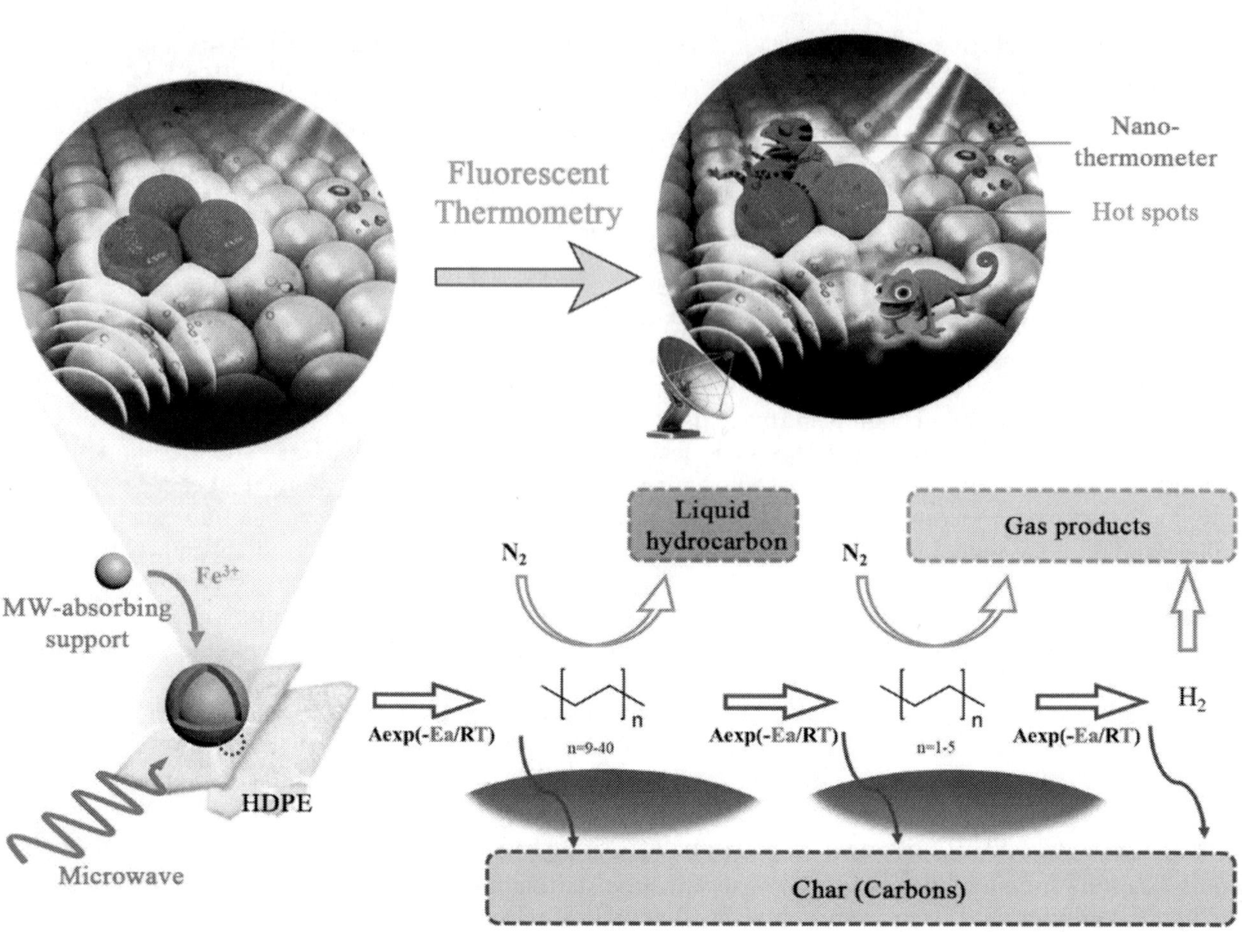
Fluorescent Thermometry
Nano-thermometer
Hot spots
Liquid hydrocarbon
Gas products
N_2
N_2
Fe^{3+}
MW-absorbing support
HDPE
Microwave
Aexp(-Ea/RT)
n=9-40
Aexp(-Ea/RT)
n=1-5
Aexp(-Ea/RT)
H_2
Char (Carbons)

298c Hydro Cracking of Waste Plastics to Naphthenes over Hierarchical Y Zeolites: From Catalyst Synthesis to Process Development

Muhammad Azam, Muhammad Usman, Ines Graca, Waheed Afzal

Polyolefins represent 36% of global plastic production and are widely used daily due to their durability, non-toxicity and chemical inertness [1]. However, the same characteristics make them the most challenging class of plastics to recycle. Globally, polyolefins account for 53% of plastic waste and even a small percentage of polyethylene (PE) gets recovered through recycling (~14% in 2015) [2,3]. Thus, the non-biodegradable and petroleum-derived unrecycled plastics result in significant economic loss and raise serious environmental concerns [4]. Also, current repurposing approach via recycling and/or downcycling has a detrimental effect on the properties of recycled materials [5]. Therefore, governments and global organisations are focusing on more effective ways to simultaneously address the ecological and economic challenges of plastics.

Photo reforming [6], depolymerization [7], deconstruction [1], gasification [8], thermal [9] or catalytic pyrolysis [9,10] and hydrocracking [11] have been exploited. However, low evolution rate (typically in mmol g^{-1}) during photo-reforming and low-value products (i.e., waxy products) during depolymerization make both processes inefficient and somewhat limited in the scope of waste plastics management. Similarly, high temperature requirements during gasification (600-700 °C) and pyrolysis (500-800 °C) and high content of olefins with products instability and prone to low-value gas products during thermal and catalytic pyrolysis lead these scenarios to some industrial issues. Contrary to that, upcycling of plastic through hydrocracking into high quality fuel significantly accumulates plastic wastes problem [12,13]. Hydrocracking enables a circular economy by utilizing plastics as a feedstock to produce fuels that mitigate the requirement for ongoing feedstock procurement and could transform significant economic losses into benefits with energy savings and reduced greenhouse gas emissions.

Hydrocracking over a bifunctional metal/acid catalysts under a reductive atmosphere is appealing since the acid support can crack carbon–carbon bonds together with the hydrogenation of intermediates over metal loading and prevent catalytic coking [14]. However, current methods and catalysts are not yet practicable for implementation due to their relatively poor yields, less acidity, low porosity and external surface area, high energy and prolonged time requirements. Therefore, it's the need of the hour to develop a catalytic technique which creates a balance between acid and metal sites for high product yields, requires low energy and tunes product distribution for various applications.

Here, commercial zeolite supports i.e., faujasite (FAU) type Y is utilized as a base catalyst. Various novel, energy-efficient, time-saving and environmentally friendly techniques have been investigated to improve the accessibility of the active sites within zeolite frameworks through hierarchical modification of zeolites. Although efficient, steaming at high temperatures is generally an energy-intensive and time-consuming process whereas acid leaching significantly produces a large quantity of acidic and toxic effluent, which limits the sustainable properties of the resulting hierarchical zeolites [15]. Also, these processes cause some noticeable changes in the acidic properties of hierarchical zeolite depending on the applied routine of dealumination, because of the adjustments in Si/Al proportion and crystallinity [16,17]. Therefore, to enhance the porosity of zeolite framework, two novel, energy-efficient and time-saving techniques are studied and results are compared with the steaming and acid-leaching approaches. The fabricated hierarchical zeolite framework retained the crystallinity, acidity and showed enhanced selectivity of liquid products in the gasoline range fuels with better affinity towards iso-paraffins and naphthalene.

Therefore, the present study highlights the catalytic hydrocracking route for the management of virgin as well as post-consumed HDPE which potentially produces renewable liquid fuels. Ni loaded zeolite Y and its respective hierarchical zeolites with enhanced porosity and external surface area were synthesized and characterized by low and wide angle XRDs, TEM, FTIR-py, DRS UV-vis, N -physiorption, and H -TPR.

Hydrocracking experiments were performed in a 300 ml stirred batch Parr reactor (series 4560 mini reactors) at 10-30 bar initial cold H_2 pressure at a temperature of 350-400 °C for a reaction time of 15-60 min and with a feed to catalyst ratio of 20:1.

Compared to the pristine Y zeolite, Ni incorporated Y zeolite showed better conversion and selectivity to oils due to hydrogenation ability, and additional Lewis acidity of Ni loaded samples. Similarly, hierarchical modification of the Y zeolite led to a considerable improvement of the conversion and selectivity to liquid products, which were further boosted to >99% by the addition of Ni. The increase in the external surface area and mesoporous volume that improved the bulk polymer molecules diffusion, combined with the uniformly dispersed small Ni particles, were at the origin of the observed behaviour. Based on the GC-analysis, all Ni loaded hierarchical Y zeolite samples showed remarkable selectivity towards lighter hydrocarbons (C_5-C_{12}) with a little affinity towards diesel range fuels which was shifted to 100 % gasoline range at 30 bar H_2 pressure. Similarly, the Ni loaded composite samples showed enhanced production of aromatics and naphthenes with reduced selectivity of n-paraffins. Compared to virgin HDPE, all the Ni loaded hierarchical Y zeolites showed similar conversion and selectivity over post-consumed HDPE milk bottles. Although, there was some decline in activity due to coke deposition and Ni sintering after multiple reaction runs, however, the zeolites performance could be almost completely recovered to their initial activity through thermal regeneration.

Moreover, the advantages and challenges of hydrocracking of real waste plastics streams over other linear end of life (i.e., incineration) and circular end of life (i.e., pyrolysis) were quantified by conducting techno economic and life cycle assessment. The study was focused on the process simulation which identify the key pathways to enhance the impact of each process for a better plastics circular economy. From the sustainable perspective, life cycle assessment and techno economic analysis showed the benefits of hydrocracking over pyrolysis and incineration.

Therefore, comparative analyses showed that produced hierarchical zeolites through novel techniques have dramatically increased catalytic activity for catalysis involving large substrates, like catalytic cracking and hydrocracking of polymers into narrower gasoline range fuels with minimum environmental impacts. Overall, the present study proposed a promising route to government organizations, industries and stakeholders to achieve 'Net Zero' plan and to execute better plastic waste management option in the long run.

References

1. Conk, R. J., Hanna, S., Shi, J. X., Yang, J., Ciccia, N. R., Qi, L., & Hartwig, J. F. Catalytic deconstruction of waste polyethylene with ethylene to form propylene. *Science* **377**, 1561–1566 (2022).
2. Geyer, R., Jambeck, J. R. & Law, K. L. Production, use, and fate of all plastics ever made. *Sci Adv* **3**, (2017).
3. Coates, G. W. & Getzler, Y. D. Y. L. Chemical recycling to monomer for an ideal, circular polymer economy. *Nature Reviews Materials* **5**, 501–516 (2020).
4. Garcia, J. M. & Robertson, M. L. The future of plastics recycling. *Science (1979)* **358**, 870–872 (2017).
5. Abel, B. A., Snyder, R. L. & Coates, G. W. Chemically recyclable thermoplastics from reversible-deactivation polymerization of cyclic acetals. *Science* **373**, 783–789 (2021).
6. Jiao, X. *et al.* Direct Polyethylene Photoreforming into Exclusive Liquid Fuel over Charge-Asymmetrical Dual Sites under Mild Conditions. Nano Letters **22**, 10066-10072, (2022)
7. Lal, A. & Chen, H. Depolymerization of HDPE to wax in the presence of a catalyst formed by homonuclear macrocyclic zirconium complex chemically bonded to alumina support. *Appl Catal A Gen* **303**, 9–17 (2006).
8. Lopez, G. *et al.* Hydrogen Production by High Density Polyethylene Steam Gasification and In-Line Volatile Reforming. *Ind Eng Chem Res* **54**, 9536–9544 (2015).
9. Yao, L., Zhu, J., Li, S., Ma, Y. & Yue, C. Analysis of liquid products and mechanism of

thermal/catalytic pyrolysis of HDPE. *J Therm Anal Calorim* **147**, 14257–14266 (2022).
10. Wang, Y. *et al.* Catalytic Pyrolysis of Polyethylene for the Selective Production of Monocyclic Aromatics over the Zinc-Loaded ZSM-5 Catalyst. *ACS Omega* **7**, 2752–2765 (2022).
11. Vela, F. J. *et al.* Enhancing the performance of a PtPd/HY catalyst for HDPE/VGO hydrocracking through zeolite desilication. *Fuel* **329**, 125392 (2022).
12. Zhang, F., Zeng, M., Yappert, R. D., Sun, J., Lee, Y. H., LaPointe, A. M., ... & Scott, S. L Polyethylene upcycling to long-chain alkyl aromatics by tandem hydrogenolysis/aromatization. *Science* **370**, (2020).
13. Rahimi, A. R. & Garciá, J. M. Chemical recycling of waste plastics for new materials production. *Nature Reviews Chemistry* **1**, 1–11 (2017).
14. Liu, S., Kots, P. A., Vance, B. C., Danielson, A. & Vlachos, D. G. Plastic waste to fuels by hydrocracking at mild conditions. *Sci Adv* **7**, 8283–8304 (2021).
15. Abdulridha, S., Zhang, R., Xu, S., Tedstone, A., Ou, X., Gong, J., ... & Fan, X. An efficient microwave- assisted chelation (MWAC) post-synthetic modification method to produce hierarchical Y zeolites. *Microporous and Mesoporous Materials* **311**, 110715 (2021).
16. Koohsaryan, E. & Anbia, M. Nanosized and hierarchical zeolites: A short review. *Chinese Journal of Catalysis* **37**, 447–467 (2016).
17. Feliczak-Guzik, A. Hierarchical zeolites: Synthesis and catalytic properties. *Microporous and Mesoporous Materials* **259**, 33–45 (2018)

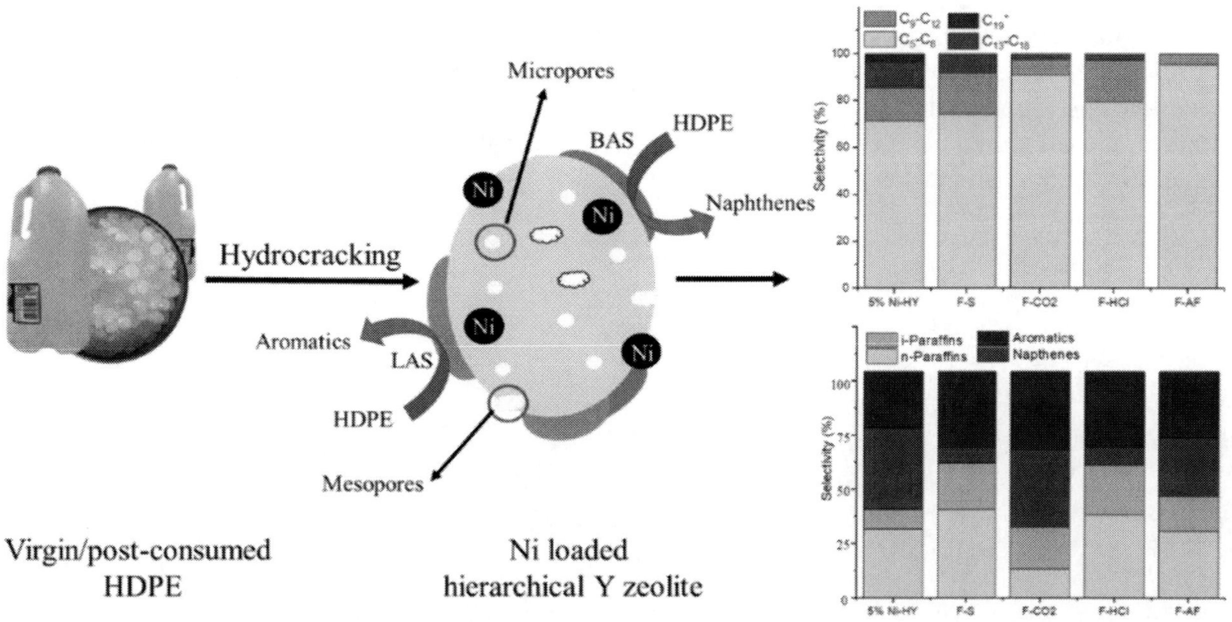

298d Co-Depolymerization of Lignin and Plastic Waste Using Hydrothermal Liquefaction Technology

Somtochukwu Anonyuo, Sampath Karunarathne, Matthew Kline, Sampath Gunukula, Clayton Wheeler

Depolymerization of plastics is one of the most challenging tasks in the chemical upcycling of plastic-based waste because the disassociation of the stable carbon-carbon bonds is only possible at a very high reaction temperature. Lignin is the second most abundant biopolymer in nature, which is currently underutilized. Therefore, it is desirable to find a sustainable solution for increasing the use of underutilized lignin and plastic beyond incineration for energy. Processes like pyrolysis and combustion are associated with energy disadvantages as there is a need to dry the feedstock before processing. However, this shortcoming can be overcome with Hydrothermal liquefaction technology (HTL) due to its ability to upgrade wet materials to value-added products like high-quality fuel. In this research, the effects of HTL process conditions (temperature between 300℃ to 400℃, pressure, and residence time) on the quality and quantity of co-depolymerization of plastics and lignin products are delineated. The dichloromethane solvent is used to extract the organics from the HTL product mixture. We further determine the effect of the lignin chemical structure and branching structure of polyolefins on the quality of produced fuel intermediates under supercritical water conditions. Gas chromatography, Nuclear Magnetic Resonance (NMR) spectroscopy, and elemental analysis results are used to propose a major reaction mechanism of lignin and polyolefin conversion to fuel intermediates under sub and supercritical conditions. The initial results show that the feedstock conversion is found to increase as the temperature increases. However, for lignin and polyethylene mixture, the oil phase yield is seen to be maximum around 370℃ and a decline in the oil yield is seen as the temperature increases beyond 370℃.

298e Comparative Techno-Economic Analysis of Producing High-Value Chemicals and Fuels from Waste Plastic Via Conventional Pyrolysis and Thermal Oxo-Degradation.

Olumide Olafasakin, Jiaze Ma, Victor M. Zavala, Mark Mba Wright

The rapid rise in global plastic production in recent decades has resulted in a massive generation of plastic waste. Over 75% of the plastic waste generated in the United States was sent to landfill, with a meager 8.7% recycled [1], [2]. Chemical recycling of plastic waste is gaining more attention, and the thermochemical decomposition of waste plastics into valuable chemicals and liquid fuel is being researched and developed. However, there are few studies on the economic feasibility of this pathway for dealing with plastic waste[2].

This study conducts a comparative techno-economic analysis on the thermochemical decomposition of waste high-density polyethylene to produce gaseous (ethylene and propylene) and liquid products (naphtha, diesel, and wax range hydrocarbons) via conventional pyrolysis (low residence time and high residence time) and thermal oxo-degradation. The study also included scenarios with pyrolysis oil upgrading.

Preliminary techno-economic analysis showed that the capital cost could range from $45.56 million for thermal oxo-degradation without hydrocracking to 134.33 for conventional pyrolysis scenarios. Scenarios with hydrocracking resulted in annual revenues in the range of $53 million to $58 million, while scenarios without hydrocracking resulted in annual revenues in the range of $21 million to $40 million. However, the net present value (NPV) showed that the most economically favorable scenarios are the scenarios with no hydrocracking with NPVs ranging from $81million to $118 compared to scenarios with hydrocracking ranging from $42million to $80 million as the high revenue could not compensate for the high capital investment. Sensitivity analysis showed that product yield and capital cost have the biggest impact on the process economics of the facilities.

The preliminary result shows that plastic pyrolysis via the different pathways can be economically feasible with scenarios with product upgrading yielding the largest annual revenues but the lowest net present value due to its high capital cost.

References

[1] D. Zhao, X. Wang, J. B. Miller, and G. W. Huber, "The Chemistry and Kinetics of Polyethylene Pyrolysis: A Process to Produce Fuels and Chemicals," *ChemSusChem*, vol. 13, no. 7, pp. 1764–1774, Apr. 2020, doi: 10.1002/CSSC.201903434.

[2] Houqian Li *et al.*, "Expanding plastics recycling technologies: chemical aspects, technology status and challenges," *Green Chemistry*, 2022, doi: 10.1039/D2GC02588D.

298f Advances in Steam Assisted Catalytic Conversion of Waste Plastics into Liquid Fuels and Chemicals

Foster Agblevor, Corey Skenandore, Ezra Johnson, Corbin Romney

- Plastics are some of the most useful materials ever evented by man. Plastics are found in consumer products, engineering products, automobiles, airplanes, marine vehicles etc. Since the introduction of synthetic organic polymeric products (resins and fibers) into the consumer market in the early 1950s, the global plastics production has risen from a mere 2 million metric tonnes (Mt) in 1950 to over 380 Mt by 2015. The largest application of plastics is in packaging where the volume has accelerated because of single use containers. Plastics content in municipal solid waste (MSW) has increased from 1% in 1960 to more than 10% by 2005. None of the commonly used plastics are biodegradable and accumulate in landfills or in the natural environment. The most prevalent plastics are: high density polyethylene (HDPE), low density and linear-low density polyethylene (LDPE), polypropylene (PP), polystyrene (PS), polyvinylchloride (PVC), polyethylene terephthalate (PET), polyurethanes (PUR), polyesters, polyamide, acrylic fibers. The pure polymers are usually mixed with additives to produce desirable properties. Despite its usefulness, petroleum derived plastics have become a major pollution problem worldwide glugging gutters, rivers, oceans and landfills. The need to convert waste plastics into useful consumer products and thus reducing environmental pollution cannot be overstated. Current, recycling methods cannot solve the pollution problems. We have investigated the conversion waste plastics (HDPE, PP, PS) into fuels and aromatic compounds. The reactions were conducted under autogeneous pressure reaction conditions in temperature range of 300 to 400 C and reaction times of 15 to 30 min. The catalyst used was a proprietary catalyst formulated in-house. The ratio of water to plastic ratio (W/P) was varied from 0.2 to 10. When W/P ratio was low, the PP reaction products were mostly aromatic compounds, but when the W/P ratio was high the reaction products were mostly linear alkanes. The products from HDPE were mostly alkanes and did not contain much aromatics. The HDPE reactions were strongly affected by the reaction temperature. Analysis of the reaction products showed that the products at high temperatures will qualify as gasoline fraction whereas at lower temperatures the diesel fractions were relatively high. The yields of the liquid products were very high and the gas yield were relatively low. There was hardly any solids production in this process.

298g Investigation of Recyclable and Renewable Crude Oil from Biomass and Plastics

Golam Chowdhury, Marco J. Castaldi

In the present world, the nonrenewable source of petroleum crude oil is suspected to cause irreversible climate impacts. So, refinery companies are looking for renewable and recyclable crude oil (RCOs). Biomass and Plastics are great source of RCOs. The goal of this work is to design, build and operate a pyrolysis reactor which is capable of converting different solid feedstock derived from waste to pyrolysis oil. The oils are characterized by GC-MS and sent to BASF Corporation in Iselin, NJ for catalytic upgrading potential. In this study, the feedstock ranged from virgin plastic resins (LDPE, HDPE, PS, PP), recycled rubber from tires to poplar wood chips. It was found that, increasing the temperature from 500 °C to 600 °C for LDPE and HDPE decrease the liquid oil yield from 61.1% to 54.7% and 79.0 wt.% to 75.5 wt.% respectively. The olefin to paraffin ratio increases as the temperature increases for LDPE and HDPE. The ratio increases on average of 1.25 times. From the other polymer resins feed stock, it was found that, the polystyrene gives higher liquid yields (86.41 wt.%). The oxygenated entities in the matrix for biomass decreases as temperature increases from 400 °C to 500 °C due to deoxygenation. The catalytic cracking showed that PE pyrolysis oil yields higher olefins than Vacuum gas oil (VGO). The catalytic cracking of LDPE pyrolysis oil results in 9.8 wt.% more naphtha yield and 1 wt.% coke yield than the VGO catalytic cracking.

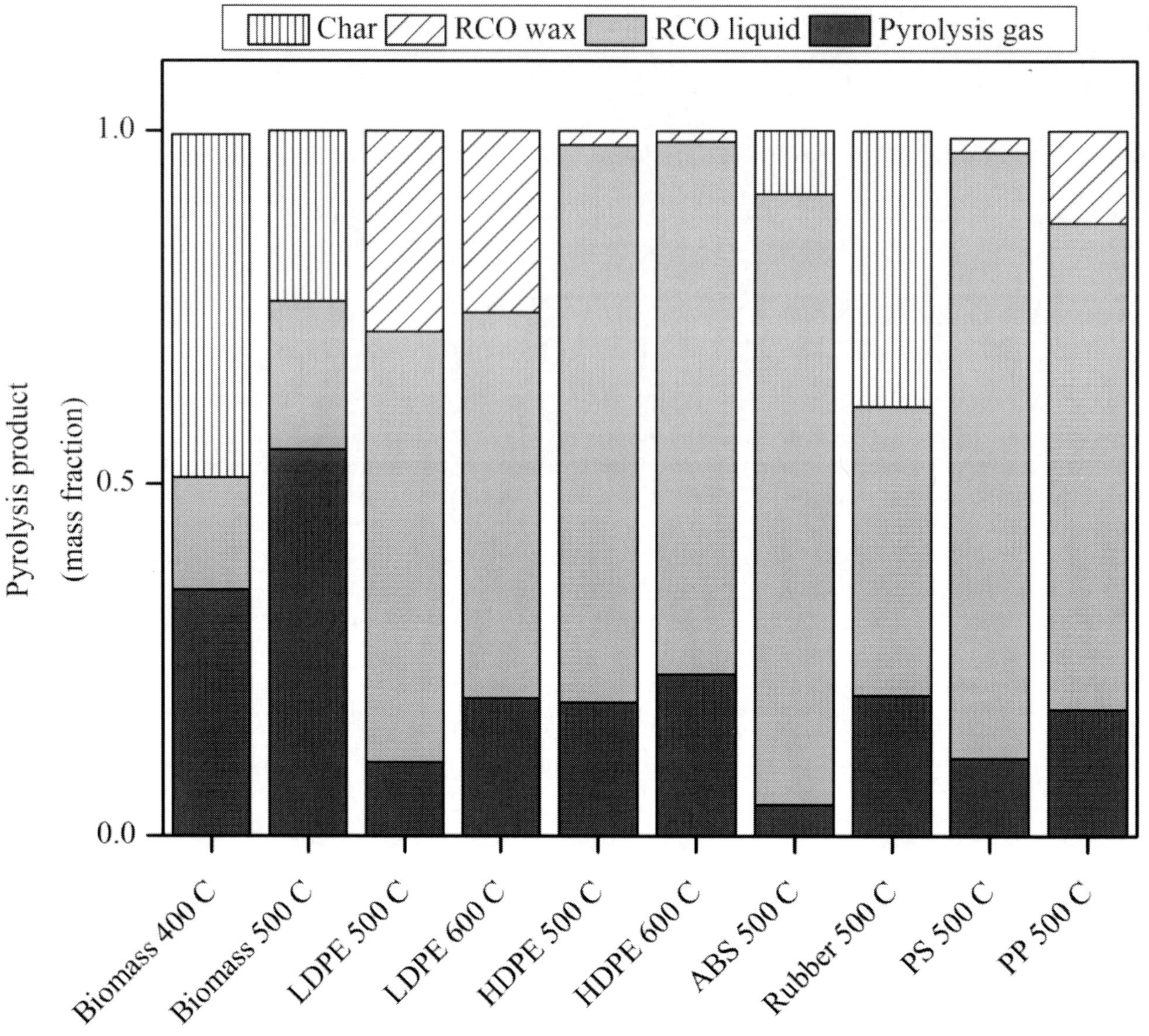
Char
RCO wax
RCO liquid
Pyrolysis gas
Pyrolysis product
(mass fraction)
1.0
0.5
0.0
Biomass 400 C
Biomass 500 C
LDPE 500 C
LDPE 600 C
HDPE 500 C
HDPE 600 C
ABS 500 C
Rubber 500 C
PS 500 C
PP 500 C

57a Plastic Policy Developments for Chemical Engineers 2023

Mary Ternes

Plastic policy and foundations for new policy continue to develop rapidly, globally, nationally and in individual states. The world is meeting to discuss solutions to the plastic waste and pollution crisis in the United Nation negotiations for the March 2, 2022 Resolution to end plastic pollution with an international legally binding instrument by 2024, with its third meeting in November, 2023. The United States Environmental Protection Agency has proposed a draft plastic recycling strategy, requesting public comment on all aspects of plastic. Individual states like California have made significant progress regarding developing a toxicity metric for microplastic exposure, sampling and monitoring protocol for microplastic pollution, banning problematic plastic uses and imposing extended producer responsibility upon plastic producers, while others have banned such approaches. This discussion will review the current status of plastic thought and developing future policy and regulation regarding plastic production, plastic waste and plastic pollution globally, in the United States and among the states including: ongoing global efforts to develop an international agreement to end plastic pollution by 2024 including the emerging hierarchy of approaches to mitigating plastic pollution as well as areas of agreement, convergence, conflict and aggressive strategic movement in this space, unpacking these issues for chemical engineers; how current plastic design and plastic waste management fails to prevent plastic pollution and renders recycling problematic; where the United States falls short in implementing delegated federal legislation and impact of environmental justice authority; range of action by individual states and potential issues arising from this current patch work quilt of state specific plastic legislation. Attendees should gain a better understanding of the potential for future regulation and insight into business risk management approaches for those entities relying significantly on plastic in their business models, whether producing plastic or simply utilizing plastic in daily business activities.

With this presentation you will:

1. Learn the current status and potential future policy approaches that may govern plastic products in order to better manage plastic waste and mitigate plastic pollution;
2. Understand the fundamental barriers posed by the current heterogenous nature of post-use plastic in reuse and recycling;
3. Appreciate the business risk posed by failure to properly manage plastic waste and prevent plastic pollution arising from recent developments in characterizing risk from plastic pollution to human health and the environment.
4. Gain insight helpful in developing new approaches that may insulate businesses from the uncertainty of future plastic regulation.

57b Scale-Up of PureCycle Solvent Dissolution Recycling Process

Dustin Olson

PureCycle uses a patented, ground-breaking recycling process to separate virtually all color, odor, and contaminants from polypropylene waste feedstock to transform it into Ultra-Pure Recycled (UPR) resin. PureCycle's UPR resin is a like-virgin material that is easily colorable, and its production is projected to consume 79% less energy and produce 35% less carbon emissions than new polypropylene manufacturing. This presentation will outline the evolution of the PureCycle process from the initial concept to commercial reality. The key challenges and successes at each stage of development will be discussed.

57c Advanced Recycling Applications around the World

Shelby Browning

Plastic waste has become a pressing global concern, necessitating the exploration of advanced recycling applications. This talk provides a comprehensive analysis of the various types of plastic recycling, and heat-based chemical recycling. The talk delves into the multifaceted challenges encountered when recycling plastics in different regions across the globe. It considers crucial factors such as infrastructure limitations, regulatory frameworks, and the role of technological advancements in shaping recycling practices. By examining these challenges, the talk aims to foster a deeper understanding of the complexities involved in implementing effective recycling systems.

Furthermore, this talk briefly explores the relationship between recycling and carbon emissions, as well as the potential community benefits associated with waste plastic recycling. It discusses how recycling initiatives can create employment opportunities, foster sustainable economic growth, and promote resource conservation. Additionally, it emphasizes the positive impact of recycling on reducing landfill usage and mitigating environmental pollution.

By providing a comprehensive overview of various recycling applications and their global significance, this talk aims to inspire and mobilize collaborative efforts among stakeholders. It emphasizes the urgent need for collective action to address the challenges posed by plastic waste and harness the immense benefits of recycling on a larger scale. Together, we can pave the way towards a more sustainable and circular future.

154af Polymer-Based Catalyst for Efficient Glycolysis of Poly (ethylene terephthalate) (PET)

Fahimeh Forouzeshfar, Maria Coleman, Joseph Lawrence

Glycolysis is one of the advanced recycling methods for Polyethylene Terephthalate (PET) that can handle feedstock not recycled by mechanical recycling. It has the potential to create a more sustainable future for PET fibers used in clothing. Glycolysis reaction involves breaking down of the PET in excess ethylene glycol into Bis (2- hydroxyethyl) terephthalate (BHET), which can further be converted into PET. Glycolysis typically requires the use of a catalyst to break down the PET polymer into its constituent monomers. Since catalysts are crucial to the glycolysis process, it is important to recover them efficiently for reuse. This study aims to develop a catalyst for glycolysis of PET that does not require extensive and expensive recovery techniques such as distillation. We developed a catalyst by combining a polymeric acid (polyacrylic acid) and a super base 1,8-diazabicyclo [5.4.0] undec-7- ene (DBU) and investigated its effectiveness and selectivity. The impact of reaction time, temperature, catalyst/PET, and EG/PET ratios on PET conversion and BHET yield were monitored. Ultrafiltration method was used to selectively separate the polymer-based catalyst from other components of the reaction mixture, in order to recover and reuse. A reaction mechanism for glycolysis of PET using polymer-based catalyst was proposed and confirmed by FTIR and NMR methods.

57e Revolutionizing Sustainability: TimePlast's Integration of Advanced Computing and Engineering for Water Soluble, Time-Programmable Bioplastics

Manuel Rendon, Tony Martinez, Thomas Anderson, Christina Fontanesi

TimePlast is at the forefront of environmental innovation, specializing in the creation of water-soluble, time-programmable bio-plastics. Utilizing cutting-edge technologies in the space of engineering simulation and AI.

TimePlast in partnership with Ecological Building & Supply L.L.C. is optimizing its unique molecular degradation process to create plastics that dissolve in water within a user-defined time frame. This collaboration of advanced computing, Generative AI and automated engineering technologies will allow the company to tailor the degradation process, ranging from minutes to years, aligning with specific industry applications. By integrating these technological advancements, TimePlast is setting new standards in sustainable materials, providing an eco-friendly alternative to conventional plastics, and positioning itself as the global leader in the fight against plastic pollution.

154a Discovery and Mechanism-Guided Engineering of BHET Hydrolases for Improved PET Recycling and Upcycling

Xiujuan Li, Anni Li, Luxuan Wu, Yibo Song, Rongrong Yang, He Huang

Plastics have facilitated human life and become an integral part of packaging, healthcare, aerospace, and agriculture due to their durability, transparency, and economic viability[1]. Due to the lack of or low activity of catabolic microorganisms/enzymes that can degrade these plastic constituents, many plastics persist significantly in terrestrial and marine environments[2]. Microplastics have even been found in human feces, blood, and placenta, indicating that plastic pollution is a threat to human health and needs more attention[2]. Polyethylene terephthalate (PET), a significant portion of plastic waste[3], is exceedingly hard to degrade due to the chemical inertness of the ester linkages and the aromatic nuclei[4]. Several PET-degrading microorganisms (e.g., *Ideonella sakaiensis*[3], *Thermobifida cellulosilytica*[5], *Saccharomonospora viridis*[6], *Fusarium solanipisi*[7], *Nocardiopsaceae*[8]) and enzymes (e.g., lipase[9], esterase[10], cutinase[7], PETase[3], and mono-2-hydroxyethyl terephthalate hydrolase (MHETase)) have been well explored[3]. However, current PET recycling without BHETase and/or relevant bacteria still faces the bottleneck that the heterogeneous products yielded from PET degradation are unfavorable for PET recondensation and high-value derivative synthesis.

In our study, Two BHETases (ChryBHETase from *Chryseobacterium sp.* PET-29 and BsEst from *Bacillus subtilis* PET-86) are identified from the environment via enzyme mining, including three phases: (i) microorganism determination; (ii) enzyme identification; (iii) *in vitro* and *in silico* characterization. After the computational study, we found that a certain number of water molecules were present in SBC, indicating the larger BHET might be blocked by uncertain barrier structures (the particular region is called lid in hydrolase [11]). Subsequently, mechanism-guided barrier engineering is employed to yield two robust and thermostable ΔBHETases with up to 3.5-fold enhanced k_{cat}/K_M than wild-type. Coupling ΔBHETase into a two-enzyme system overcomes the challenge of heterogeneous product formation and results in up to 7.0-fold improved TPA production than seven state-of-the-art PET hydrolases (PETase, DepoPETase, FAST-PETase, ThermoPETase, DuraPETase, LCC, and LCC-ICCG), under the conditions used here. Finally, we employ a ΔBHETase-joined tandem chemical-enzymatic approach to valorize 21 commercial post-consumed plastics into virgin PET and an example chemical (*p*-phthaloyl chloride) for achieving the closed-loop PET recycling and open-loop PET upcycling.

Reference

(1) Borrelle, S. B.; Ringma, J.; Law, K. L.; Monnahan, C. C.; Lebreton, L.; McGivern, A.; Murphy, E.; Jambeck, J.; Leonard, G. H.; Hilleary, M. A.; Eriksen, M.; Possingham, H. P.; De Frond, H.; Gerber, L. R.; Polidoro, B.; Tahir, A.; Bernard, M.; Mallos, N.; Barnes, M.; Rochman, C. M. Predicted Growth in Plastic Waste Exceeds Efforts to Mitigate Plastic Pollution. Science 2020, 369 (6510), 1515–1518. https://doi.org/10.1126/science.aba3656.

(2) Yong, C. Q. Y.; Valiyaveettil, S.; Tang, B. L. Toxicity of Microplastics and Nanoplastics in Mammalian Systems. Int. J. Environ. Res. Public Health 2020, 17 (5), 1509. https://doi.org/10.3390/ijerph17051509.

(3) Yoshida, S.; Hiraga, K.; Takehana, T.; Taniguchi, I.; Yamaji, H.; Maeda, Y.; Toyohara, K.; Miyamoto, K.; Kimura, Y.; Oda, K. A Bacterium That Degrades and Assimilates Poly(Ethylene Terephthalate). Science 2016, 351 (6278), 1196–1199. https://doi.org/10/bc8g.

(4) Feng, B.; Jing, Y.; Guo, Y.; Liu, X.; Wang, Y. Plastic Waste to Drug Intermediate: Targeted Cleavage of C–O Bonds in Polyphenylene Oxide to 3,5-Dimethyl Phenol. Green Chem. 2021, 23 (23), 9640–9645. https://doi.org/10.1039/D1GC02767K.

(5) Herrero Acero, E.; Ribitsch, D.; Steinkellner, G.; Gruber, K.; Greimel, K.; Eiteljoerg, I.; Trotscha, E.; Wei, R.; Zimmermann, W.; Zinn, M.; Cavaco-Paulo, A.; Freddi, G.; Schwab, H.; Guebitz, G. Enzymatic Surface Hydrolysis of PET: Effect of Structural Diversity on Kinetic Properties of Cutinases from Thermobifida. Macromol. 2011, 44 (12), 4632–4640. https://doi.org/10.1021/ma200949p.

(6) Kawai, F.; Oda, M.; Tamashiro, T.; Waku, T.; Tanaka, N.; Yamamoto, M.; Mizushima, H.; Miyakawa, T.; Tanokura, M. A Novel Ca2+-Activated, Thermostabilized Polyesterase Capable of Hydrolyzing Polyethylene Terephthalate from Saccharomonospora Viridis AHK190. Appl. Microbiol. Biotechnol. 2014, 98 (24), 10053–10064. https://doi.org/10.1007/s00253-014-5860-y.

(7) Eberl, A.; Heumann, S.; Brückner, T.; Araujo, R.; Cavaco-Paulo, A.; Kaufmann, F.; Kroutil, W.; Guebitz, G. M. Enzymatic Surface Hydrolysis of Poly(Ethylene Terephthalate) and Bis(Benzoyloxyethyl) Terephthalate by Lipase and Cutinase in the Presence of Surface Active Molecules. J. Biotechnol. 2009, 143 (3), 207–212. https://doi.org/10.1016/j.jbiotec.2009.07.008.

(8) Liu, Y.; Liu, C.; Liu, H.; Zeng, Q.; Tian, X.; Long, L.; Yang, J. Catalytic Features and Thermal Adaptation Mechanisms of a Deep Sea Bacterial Cutinase-Type Poly(Ethylene Terephthalate) Hydrolase. Frontiers in Bioengineering and Biotechnology 2022, 10.

(9) Tan, Y.; Henehan, G. T.; Kinsella, G. K.; Ryan, B. J. An Extracellular Lipase from Amycolatopsis Mediterannei Is a Cutinase with Plastic Degrading Activity. Comput. Struct. Biotechnol. J. 2021, 19, 869–879. https://doi.org/10.1016/j.csbj.2021.01.019.

(10) Ribitsch, D.; Herrero Acero, E.; Greimel, K.; Dellacher, A.; Zitzenbacher, S.; Marold, A.; Rodriguez, R. D.; Steinkellner, G.; Gruber, K.; Schwab, H.; Guebitz, G. M. A New Esterase from Thermobifida Halotolerans Hydrolyses Polyethylene Terephthalate (PET) and Polylactic Acid (PLA). Polymers 2012, 4 (1), 617–629. https://doi.org/10.3390/polym4010617.

(11) Gricajeva, A.; Nadda, A. K.; Gudiukaite, R. Insights into Polyester Plastic Biodegradation by Carboxyl Ester Hydrolases. J. Chem. Technol. Biotechnol. 2022, 97 (2), 359–380. https://doi.org/10.1002/jctb.6745.

Figure. The mechanism-guided barrier engineering yielded the improved ΔBHETase. (a) Schematic representation of the mechanism-guided barrier engineering strategy. Step 1: *In silico* prediction of structure and solvation observables. Step 2: Design truncated variants. Step 3: Statistical normative scores by assigning weights of 10%, 30%, 40%, and 20% to the binding energy, number of hydrogen bonds, serine affinity attack distance, and cavity volume, respectively. Step 4: *In vitro* validation.

a

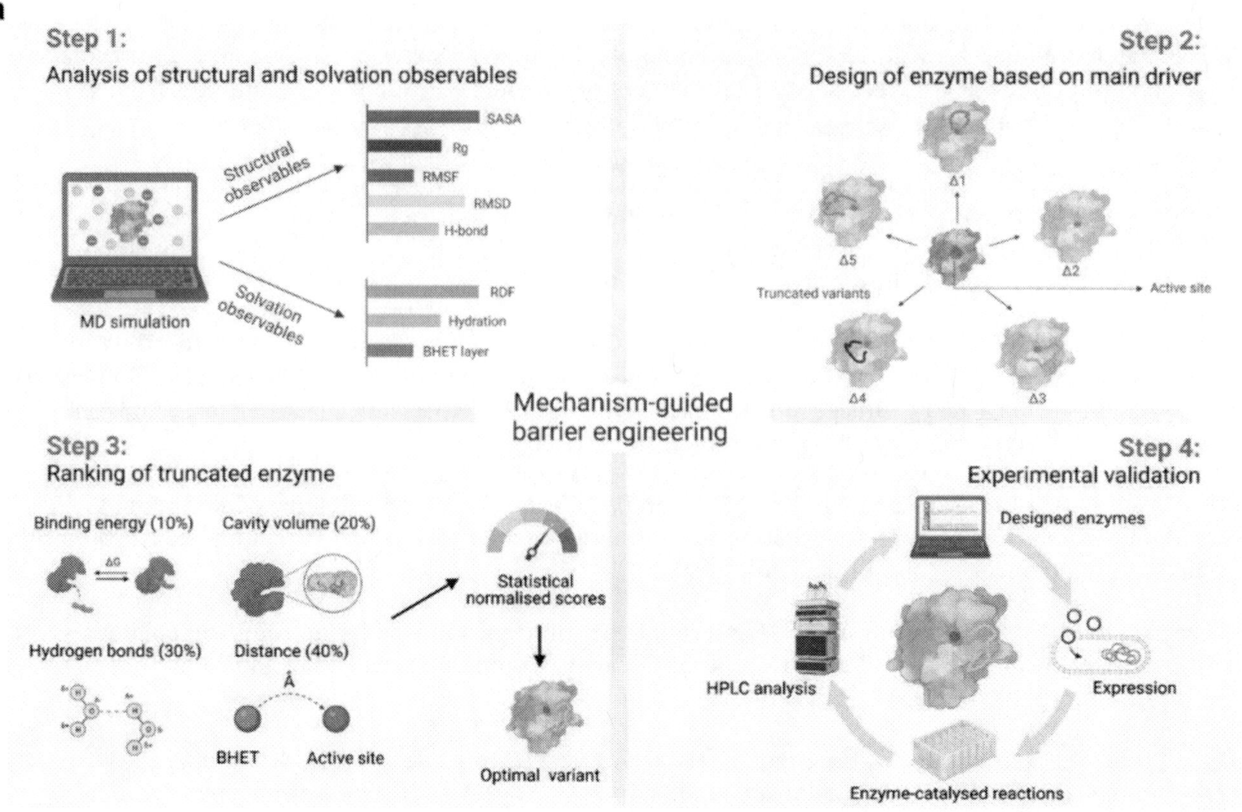

Step 1:
Analysis of structural and solvation observables
MD simulation
Structural observables
Solvation observables
SASA
Rg
RMSF
RMSD
H-bond
RDF
Hydration
BHET layer
Step 2:
Design of enzyme based on main driver
Δ1
Δ2
Δ3
Δ4
Δ5
Truncated variants
Active site
Mechanism-guided
barrier engineering
Step 3:
Ranking of truncated enzyme
Binding energy (10%)
Cavity volume (20%)
ΔG
Hydrogen bonds (30%)
Distance (40%)
Å
BHET
Active site
Statistical
normalised scores
Optimal variant
Step 4:
Experimental validation
Designed enzymes
HPLC analysis
Expression
Enzyme-catalysed reactions

154b Computational High-Throughput Screening of Glycerol-Derived Solvents for Plastic Waste Recycling

Ademola Soyemi, Tibor Szilvasi

Although plastics are essential in modern life, their use results in significant waste that ends up in landfills or in the natural environment. To combat the effects of plastic pollution, efforts are being amplified on the regulatory and technological levels to increase plastic waste recycling. Solvent-based recycling, whereby a solvent selectively dissolves plastic material which can then be recovered and reprocessed into upcycled products, represents a promising method to scale up plastic recycling. Most known applications of solvent-based recycling disclosed in industrial patents and in academic literature, however, make use common solvents which may pose potential health and environmental risks. Therefore, it is crucial to develop 'green' solvents for solvent-based plastic waste recycling. For this purpose, we propose the functionalization of glycerol, a natural product from biomass as well as a byproduct of biodiesel production, that can produce new and potentially less harmful solvents capable of dissolving plastic waste. In this work, we have developed a computational methodology based on a combination of Density Functional Theory, machine learning and Hansen theory which allows for the high-throughput screening of thousands of solvent candidates for plastics commonly found in waste streams. Consequently, we have developed a database of ~10,000 glycerol-based solvents and calculated their physical properties such as the dipole moment, density, melting point, LogS, LogP, Hansen Solubility Parameters (HSPs), and based on the HSPs of the solvents and selected plastics in addition to the melting point of the solvents, we give recommendations for solvents that can carry out selective plastic dissolution for complex plastic waste feedstock.

Reference: *Ind. Eng. Chem. Res.*, 2023. accepted DOI: 10.1021/acs.iecr.2c04567.

154c Novel Thin Film Composite (TFC) Membranes Fabricated from Upcycled Waste PVC Pipe Support Layer for Water Treatment

Atta Ur Razzaq, Milad Esfahani

Nanofiltration (NF) thin-film-composite (TFC) membranes are well-known energy-efficient membranes for water treatment. Generally, these membranes are fabricated on support layer made of commercially available materials such as polyethersulfone (PES). On the other hand, plastic pollution is one of the main environmental concerns. Therefore, it is beneficial to upcycle waste plastic to a high-added-value product such as a membrane. In this work, we demonstrate the successful fabrication of the thin film composite (TFC) membranes with the support layer made from upcycled waste PVC pipe. The ultrafiltration (UF) support membranes were fabricated from waste PVC pipe (22 wt.% polymer in a mixture of N,N-Dimethylformamide and Tetrahydrofuran 3:2 (v/v)) via nonsolvent induced phase separation (NIPS) and then a polyamide layer was synthesized on the upcycled PVC membrane using interfacial polymerization. The physicochemical properties of the upcycled PVC membrane were compared and benchmarked with membranes fabricated using research-grade research (RG) PVC and commercial PES support layer. The results indicated that upcycled PVC pipe TFC membrane exhibited similar hydrophilicity (water contact angle around 30°), slightly more negative surface charge, and higher water permeability at 100 psi (17.2 LMH/bar) compared to RG TFC (14.5 LMH/bar) and PES TFC membranes (12.7 LMH/bar). Salt rejection of waste PVC pipe TFC membrane was also comparable to RG TFC and PES TFC as it achieved 88.9 % rejection for $MgSO_4$, 77.8 % rejection for $CaCl_2$, and 46.9% rejection for NaCl. Since the upcycled waste PVC pipe TFC membranes showed similar physiochemical properties and performance to the commercial RG PVC and PES TFC membranes, this shows that upcycling waste PVC pipe to membranes for water treatment can solve two biggest global challenges of plastic pollution and water treatment/purification simultaneously.

154e Nutrient Sourcing in Optimization of Microbial Plastic Waste Valorization

Kimia Noroozi, Laura R. Jarboe

Central Question

Plastic waste has been a long standing environmental and societal issue for the modern world, and the subject of plastic recycling is extensively studied. Biological upcycling of plastic waste is an attractive topic and the prospects of using microbial cell factories include sustainability, circular economy, and a potential for production of value-added products. While a variety of organisms have been studied for their ability to utilize plastic waste, the complexity and stability of molecules present in plastic extend the process time and decrease its yield. The oleaginous yeast *Candida maltosa* is one of the species that has been studied for its ability to utilize representative hydrocarbon molecules. While *C. maltosa*'s impressive ability to grow on hydrophobic substrates as carbon sources introduces the possibility of biological upcycling plastic waste, the efficiency of the process needs to be targeted to make it a viable option for scale up. To this end, it is essential to look into media components and their effects in the uptake process, while strategizing nutrient sourcing to enhance performance by replacing conventional media components with those that have a more pronounced impact on the yield.

Results and Conclusions

In this study, various organic and inorganic nitrogen sources including ammonium sulfate, urea and casamino acids were tested to determine their effects on biomass production when hydrocarbon molecules are used as the sole carbon source in the media. Yeast cells are equipped with a nitrogen catabolic repression system that senses the type of nitrogenous compounds present in the environment and use them according to cells selectivity. Ammonium sulfate is expected that it would result in the highest growth possible. Urea, which is an organic nitrogen source and the main component of urine can be used to replace ammonium sulfate in this process without sacrificing the yield. In addition to supporting growth, urine is a waste stream that can be effectively utilized in this process. Furthermore, when ammonium sulfate was replaced in the media with casamino acids, a 2-fold increase in final biomass production was observed. Hence, a replacement of nitrogen source is considered an effective way to increase the uptake. Phosphorus and other components of the media were also studied for replacement and optimization. Different phosphorous compounds were also tested at various concentrations and prove not to have an effect. Also, all the vitamins and trace metals in standard media were replaced by a human multivitamin which resulted in an increased growth, while significantly reducing process costs.

154f Two Enzyme Whole Cell Biocatalysts for Complete PET Degradation

Siddhant Gulati, Eric Imhoff, Qing Sun

A promising solution to the challenge of polyethylene terephthalate (PET) upcycling for environmental sustainability is enzymatic degradation.[1] Although much effort has been made towards efficient plastic degradation through protein engineering[2] and whole cell surface display[3], a systematic approach using an engineered bacterial community for complete PET degradation is still required. The hypothesis is that a bacterial community with surface displayed PET-degrading enzymes will completely degrade PET to its constituent monomers terephthalic acid (TPA) and ethylene glycol (EG). Whole cell biocatalysts displaying PET-degrading enzymes FAST-PETase[2] and MHETase[4] on the surface of *Escherichia coli (E. coli)* were constructed. Additionally, high throughput assays including bis(2-hydroxyethyl) terephthalate (BHET)-agar plates and fluorogenic fluorescein dibenzoate (FDBz)[5] were explored.

FAST-PETase and MHETase were surface displayed on *E. coli* using transporter proteins YeeJ[3] and AIDA-I[6] because of their demonstrated ability to efficiently surface display heterologous passenger proteins.[7] These whole-cell biocatalysts eliminate the need for costly enzyme purification and enhance enzymatic activity. After expression, activity of surface displayed FAST-PETase was determined by incubating cells with PET films for one week at 25°C followed by HPLC analysis to detect degradation products. Although loss of cell viability within 48 hours of expression was observed for AIDA-I-FAST-PETase, PET degradation products were detected, demonstrating the active expression of FAST-PETase. A similar procedure utilizing dissolved MHET as reactant was performed to confirm the activity of both AIDA-I-MHETase and YeeJ-MHETase via HPLC.

High-throughput screening and kinetic assays for PETase were utilized for rapid screening of PET-degrading activity. BHET plates showed visual changes in both purified and whole-cell surface displayed FAST-PETase while none was detected for MHETase, as expected. Furthermore, a fluorogenic compound with PET-like ester bonds, FDBz, was used to test PETase activity. A linear correlation between fluorescence intensity and enzyme concentration was observed with purified enzymes.

References:

(1) Yoshida, S.; Hiraga, K.; Takehana, T.; Taniguchi, I.; Yamaji, H.; Maeda, Y.; Toyohara, K.; Miyamoto, K.; Kimura, Y.; Oda, K. A bacterium that degrades and assimilates poly(ethylene terephthalate). *Science* **2016**, *351* (6278), 1196-1199. DOI: 10.1126/science.aad6359.

(2) Lu, H.; Diaz, D. J.; Czarnecki, N. J.; Zhu, C.; Kim, W.; Shroff, R.; Acosta, D. J.; Alexander, B. R.; Cole, H. O.; Zhang, Y.; et al. Machine learning-aided engineering of hydrolases for PET depolymerization. *Nature* **2022**, *604* (7907), 662-667. DOI: 10.1038/s41586-022-04599-z.

(3) Gercke, D.; Furtmann, C.; Tozakidis, I. E. P.; Jose, J. Highly Crystalline Post-Consumer PET Waste Hydrolysis by Surface Displayed PETase Using a Bacterial Whole-Cell Biocatalyst. *ChemCatChem* **2021**, *13* (15), 3479-3489. DOI: 10.1002/cctc.202100443.

(4) Knott, B. C.; Erickson, E.; Allen, M. D.; Gado, J. E.; Graham, R.; Kearns, F. L.; Pardo, I.; Topuzlu, E.; Anderson, J. J.; Austin, H. P.; et al. Characterization and engineering of a two-enzyme

system for plastics depolymerization. *Proceedings of the National Academy of Sciences* **2020**, *117* (41), 25476-25485. DOI: 10.1073/pnas.2006753117.

(5) Qiao, Y.; Hu, R.; Chen, D.; Wang, L.; Wang, Z.; Yu, H.; Fu, Y.; Li, C.; Dong, Z.; Weng, Y.-X.; et al. Fluorescence-activated droplet sorting of PET degrading microorganisms. *Journal of Hazardous Materials* **2022**, *424*, 127417. DOI: 10.1016/j.jhazmat.2021.127417.

(6) Nicchi, S.; Giuliani, M.; Giusti, F.; Pancotto, L.; Maione, D.; Delany, I.; Galeotti, C. L.; Brettoni, C. Decorating the surface of Escherichia coli with bacterial lipoproteins: a comparative analysis of different display systems. *Microbial Cell Factories* **2021**, *20* (1), 33. DOI: 10.1186/s12934-021-01528-z.

(7) Jose, J.; Meyer Thomas, F. The Autodisplay Story, from Discovery to Biotechnical and Biomedical Applications. *Microbiology and Molecular Biology Reviews* **2007**, *71* (4), 600-619. DOI: 10.1128/MMBR.00011-07.

154g Waste Plastics Valorization through Pyrolysis Processes: A System Analysis Framework

Sam Barker, Mason Aksomitus, Amir Akbari, Gary Aurand, Rui Shi

Plastic waste disposal has caused significant environmental problems. Chemical recycling is a promising solution to complement mechanical recycling, with fast pyrolysis being an effective method to handle hard-to-recycle plastics while producing valuable products such as oil, waxes, and aromatics. This study focuses on designing and simulating an open-loop recycling process for waste polypropylene (PP) through fast pyrolysis to evaluate its economic and environmental performances.

The reactor was designed to operate at 510 °C, optimizing the production of paraffin waxes and suppressing the formation of liquid petroleum gas (LPG) and aromatics. The subsequent produced waxes were further cracked using a fluid catalytic cracking (FCC) reactor in a second fluidized bed reactor, followed by three distillation columns to separate product streams and produce 99% purity gasoline and diesel. Techno-economic analysis and life-cycle assessment were conducted to evaluate the economic feasibility and environmental implications of the proposed technology, and sensitivity analysis on important process parameters was performed to obtain the optimum plant design conditions.

The results indicated that the conversion of waste PP into liquid fuels using fast pyrolysis is a sustainable and economically feasible approach that can help reduce plastic waste pollution, oil usage, and carbon dioxide emissions, achieving circular economy objectives. In conclusion, this study demonstrates the potential of fast pyrolysis as a sustainable approach to complement mechanical recycling and facilitate the transition towards a circular economy for plastic waste.

154h Catalytic Conversion of Waste PET into Bhet over Mn-Substituted $MgAl_2O_4$ Spinel As a Reusable and Regenerable Heterogeneous Catalyst

Jongin Choi, Haoxiang Zhang, Do-Young Hong

Catalytic glycolysis with ethylene glycol (EG) is a promising chemical recycling method that produces monomer-level PET with lower CO_2 emissions than other methods and various catalysts have been investigated. In the glycolysis reaction, metal oxides as heterogeneous catalysts have several advantages for separation, preparation cost, and reusability with regeneration by rinsing or thermal treatment. However, the use of a high concentration of EG to achieve high selectivity to BHET monomer in the PET glycolysis facilitates the deposition of heavy residue on the catalyst and dissolution of metal oxide into the solvent at the high reaction temperature, which leads to permanent loss of catalytic activity and difficulty of product purification by the metal contaminants. Therefore, it is indispensable to develop not only a highly stable catalyst in the reaction condition but also a recyclable and regenerable catalyst.

Spinel oxides are one of the representative materials that have excellent thermal and chemical stability and the acidic and basic properties of the materials can be tuned by partial substitution with various metal cations. Taking into account the characteristics of spinel oxide, Mn-substituted $MgAl_2O_4$ spinel oxides ($Mn_xMg_{1-x}Al_2O_4$, x= 0-1.0) were successfully synthesized by co-precipitation and applied to PET glycolysis as the robust catalyst. Mn^{2+} could substitute the Mg^{2+} in the $MgAl_2O_4$ spinel structure up to 25% without the formation of free Mn oxide which was confirmed by the increase of lattice parameter (a) from 0.8063 nm to 0.8096 nm. At higher Mn contents ($x \geq 0.3$), isolated Mn_2O_3 was detected and the lattice parameters were constant at 0.8103 nm. The PET glycolysis was conducted at 190 °C for 3 h in reflux conditions. The conversion of PET over the $Mn_xMg_{1-x}Al_2O_4$ catalyst is monotonically increased from 37.1% (x=0) to 100% (x = 0.25) and the selectivity to BHET was achieved to 95.9%. The concentration of oxygen defects generating both moderate acidic and basic sites in the $MgAl_2O_4$ spinel increased with the increase of substituted Mn content which corresponded to the enhancement of the catalytic results. The recyclability and regenerability of Mn-substituted $MgAl_2O_4$ were also conducted.

154i Blue Hydrogen Production from Waste Tire: Technical and Economic Assessment

Usama Ahmed

According to the recent research, the annual production of new car tires is over 1.6 billion which in turn generates more than 1 billion waste-tires annually (Mohajerani, A., et al. 2020). Waste tires are collected in huge quantities and transported to either landfill sites or stockpiled which accounts for more than 4 billion tires worldwide (Subramanian et al. 2020). Waste tires contains rubber which is not easily biodegradable even after the long period of time and reserves the tendency to leach toxic chemicals (Labaki, M. and Jeguirim, M. 2017) . The waste management data of US shows that 35% of the waste tires are recycled, 50% is converted to tire derived fuel (TDF) and 15% is left to decompose in the landfills or burned (Hasan, A. and I. Dincer. 2019). Thermochemical recycling of waste tires to produce energy and fuels is one of the attractive options to reduce waste while meeting energy needs. Hydrogen is a clean fuel that could be produced via gasification of waste tires followed by syngas processing. In this study, two process models are developed to evaluate the hydrogen production potential from waste tires. Case 1 involves three main processes which include the steam gasification of waste tires, water gas shift, and acid gas removal to produce the hydrogen. On the other hand, case 2 represents the integration of waste tire gasification with the natural gas reforming technology where the energy from the gasifier-derived syngas can be used to provide sufficient heat to the steam methane reforming (SMR) unit. Both models are also analyzed in terms of syngas compositions, hydrogen production rate, hydrogen purity, overall process efficiency, greenhouse gas (GHG) emissions, and hydrogen production cost. The results revealed that case 2 produced the syngas with 55% higher heating value, 28% higher hydrogen production, 7% higher H_2 purity, and 26% lower CO_2 emissions compared to case 1. The results revealed that case 2 offers 10.4% higher process efficiency and 28.5% lower H_2 production costs compared to case 1. Overall, case 2 design has been found to be more efficient and cost-effective compared to the base case design.

Keywords: Gasification, Reforming, Waste Tire, H_2 Production.

References

1. Hasan, A. and I. Dincer, 2019. Comparative assessment of various gasification fuels with waste tires for hydrogen production. International Journal Of Hydrogen Energy, 44(34): p. 18818-18826.

2. Labaki, M. and M. Jeguirim, 2017. Thermochemical conversion of waste tyres—a review. Environmental Science and Pollution Research, 24(11): p. 9962-9992.

3. Mohajerani, A., et al., 2020. Recycling waste rubber tyres in construction materials and associated environmental considerations: A review. Resources, Conservation and Recycling, 155, p. 104679.

4. Subramanian, A.S., Gundersen, T. and Adams II, T.A., 2020. Technoeconomic analysis of a waste tire to liquefied synthetic natural gas (SNG) energy system. Energy, 205, p.117830.

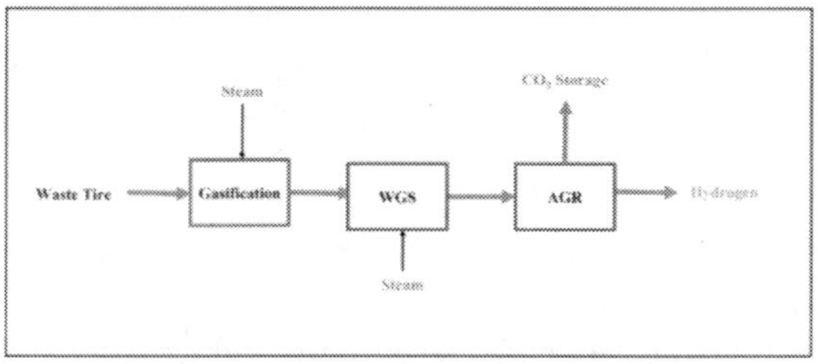

Figure 1: Case 1 base flow diagram

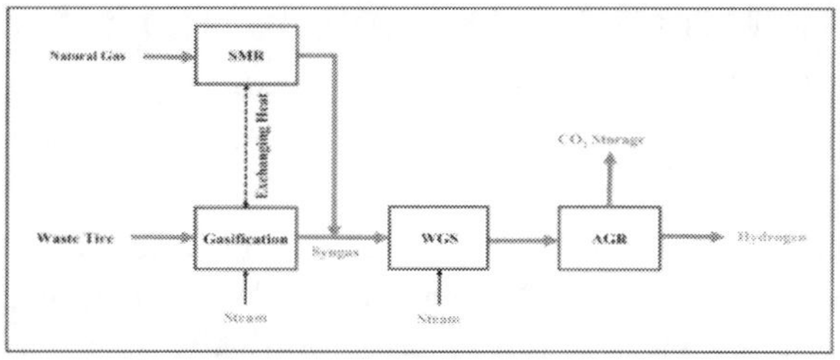

Figure 2: Case 2 base flow diagram

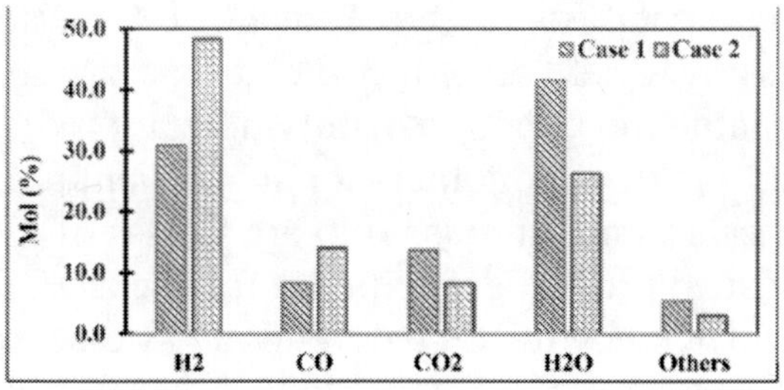

Figure 3: Syngas Composition for case 1 and case 2

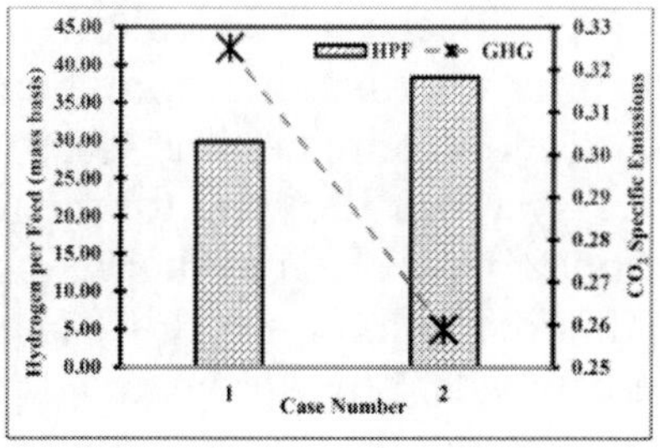

Figure 4: Comparison between case 1 and case 2 in terms of CO_2 specific emissions and normalized hydrogen production (HPF)

154j Selective and Efficient Glycolysis of Textile PET, Colored PET, and Multilayer PET Using Ionic Liquids

Fahimeh Forouzeshfar, Maria Coleman, Joseph Lawrence

Chemical recycling of polyethylene terephthalate (PET) has gained significant attention due to the increasing concern over plastic waste pollution and the need for sustainable waste management practices. Among the various methods of advanced PET recycling, glycolysis has emerged as a promising approach. The importance of chemical recycling of PET by glycolysis lies in its potential to address the limitations of traditional mechanical recycling methods, such as contamination, color mismatch, and reduced mechanical properties. The role of catalysts in glycolysis of PET is critical, as they significantly improve the depolymerization process by accelerating the rate of reaction and reducing the reaction temperature. Catalysts can also enable the selective depolymerization of complex PET systems, allowing for the recovery of high-quality monomers for subsequent reuse in the production of new PET or other value-added chemicals. Ionic liquids (ILs) are a class of catalysts that can be tailored to have a specific chemical and physical properties including for use as recoverable catalysts. This study aimed to evaluate the efficiency of ILs based upon 1,8-diazabicyclo [5.4.0] undec-7-ene (DBU) in glycolysis of PET complex systems, including textile, colored, and multilayer PET. The approach involved depolymerizing the PET using glycolysis, with the assistance of the ILs. The resulting products were purified and re-polymerized to generate new PET. The study investigated the selectivity of ionic liquid catalysts towards the target reaction in production of the desired products and unwanted side products.

154k Kinetics of Polyurethane Foam Acidolysis with Carboxylic Acids

Zach Westman, Baoyuan Liu, Kelsey Richardson, Dingyuan Lim, Alan Stottlemyer, Christopher Letko, Nasim Hooshyar, Phillip Christopher, Mahdi Abu-Omar, Vojtech Vlcek

Polyurethane (PU) is the 6th most produced polymer, with a global market size of $37.8 billion in 2020. Unlike polyolefins, polyurethanes contain reactive C-O and C-N bonds which permit facile chemical decomposition to monomeric units (polyol and isocyanate) for use in the synthesis of new PU materials. Reaction of PU materials with organic reagents such as glycols, acids, or amines has shown promise in recovering polyol products; however, each has drawbacks which have thus far prevented their widespread use. Acidolysis, the reaction of PU with organic carboxylic acids, offers polyol recovery under mild temperatures (< 200 °C) and atmospheric pressure, but its kinetics, mechanism, and scalability are not well understood. In this work, the kinetics of acidolysis of a flexible polyurethane foam with mono- and dicarboxylic acids are explored via real-time monitoring of gas (CO_2) emissions. Through a combination of kinetic measurements, analysis, and theoretical models, insights are provided into the role of acid structure on rate of acidolysis. The viability of this reaction in a real-world application is validated with kinetics of post-consumer polyurethane mattress waste. Our findings offer insights into the mechanism and kinetics of PU acidolysis and progress towards circularity of PU materials.

154m Polyethylene Terephthalate Depolymerization: Controlled Glycolysis to Oligomers *Via* Microwave Irradiation Using Antimony (III) Oxide

Somayeh Mohammadi, Martin Bouldo, Mojtaba Enayati

The production of oligomers from the glycolysis of polyethylene terephthalate (PET), instead of its fully glycolyzed product, bis-(hydroxyethyl) terephthalate (BHET) can be done under microwave using a suitable catalyst. We show that zinc acetate catalysts can generate BHET as the ultimate glycolysis product and the antimony (III) oxide (Sb_2O_3) catalyst can promote the production of higher oligomers of PET under microwave irradiation. Waste PET from drinking water bottles was glycolyzed by antimony (III) oxide (Sb_2O_3) with a loading of 0.25 wt% as the catalyst at 240°C in a Mars 6 microwave from CEM corporation, resulting in an oligomer yield of 96.7%. The conversion of PET to oligomers reached 100% in only 5 min at 240°C (with a 10 min ramping time) and a PET to ethylene glycol weight ratio of 2.5. In comparison, 0.025 wt% of zinc acetate, a well-known catalyst for PET glycolysis, produces only BHET monomer with a 96.3% yield, under the same reaction conditions. The reaction mixture and isolated products were characterized by FTIR, NMR, DSC, TGA, and HPLC. Our results demonstrated that by using Sb_2O_3 as a catalyst under microwave irradiation, the PET glycolysis can be controlled to produce higher PET oligomers as an alternative for a complete chemical depolymerization to BHET monomer. These oligomers can have potential applications as functional monomers and additives for other polymer products and also to produce high-quality second-generation products, including regenerated PET.

154o Depolymerizing PET Via 'Imidazolysis': A New Cleavage Strategy to Obtain Versatile Intermediates

Mousumi Rani Bepari, Kathryn O'Harra, Lauren Sullivan, Jason Bara

The overgrowing plastic accumulation on earth has driven scientists to come up with long-term, sustainable solutions to this issue. Chemical recycling of plastics is considered to be the most effective approach to breaking down plastics and obtaining new products. The increasing manufacturing of polyethylene terephthalate (PET), a synthetic polyester commonly used as beverage bottles, is receiving significant interest for upcycling to higher value products. Herein, we demonstrate the chemical cleavage ('olysis') of PET with a range of imidazole compounds directly in the melt phase. Through imidazolysis of PET, a diverse set of imidazoyl carbamates has been obtained, which can be further reacted to form polymers or small molecules. We outline the reaction mechanisms and conditions, and how some of these compounds can be used.

154p Molecular Recycling of Mixed Plastic Waste Facilitated By Solvents

Christian Ferger, Jacob Licht, Ali Ghasemi, Paschalis Alexandridis, Marina Tsianou

Plastics are widely used across industries such as food, beverage, textile, healthcare, automotive, aerospace, construction, agriculture, and petroleum. Only a small fraction of the plastics produced is being recycled, with the great majority landfilled or released into the environment [DOI: 10.3390/recycling6040069]. There is great interest nowadays in increasing the recycling of plastic waste. Mechanical recycling is currently used to recycle plastic waste, however, this method is efficient only for homogeneous and non-contaminated plastic waste, and for easily identifiable objects such as bottles made of PET or HDPE [DOI: 10.3390/recycling7020011]. A solution for recycling many of the plastics which mechanical recycling fails to process comes from chemical recycling, which typically involves the deconstruction of polymers to monomers or to intermediates that can be used as chemical feedstock or fuel [DOI: 10.1016/j.scp.2022.100683]. These, however, result in undesirable greenhouse gas (GHG) emissions. Our research advances solvent-based "molecular" recycling, whereby polymers are selectively dissolved and precipitated to achieve separation and recovery. Because the polymer chains do not break, this presents a promising, low-GHG methodology for recycling waste plastic [DOI: 10.1002/anie.201915651]. Through the dissolution/precipitation method, desirable types of polymers can be separated from mixtures or blends, purified from additives or fillers, and obtained in a form suitable for reprocessing, without negatively affecting the properties of the original polymers. Examples are presented on polyolefin recovery from dissolution/precipitation using environmentally responsible solvents, and processing conditions for efficient recovery are discussed.

154q Material Recycling of Acrylonitrile Butadiene Styrene from Toy Waste Using Safer Solvents

Taofeng Lu, Wan-Ting Chen

Toy waste is an issue for municipal plastic recycling programs. In particular, toy waste typically is composed of different plastics and some metals. Separation of different parts of toy waste is necessary before recycling. In this study, a dissolution-precipitation process coupled with density separation was developed to extract acrylonitrile butadiene styrene (ABS) from toy waste using safer solvents and antisolvents. The polymer solubility of ABS was measured using Hansen solubility parameters, which were then used to down-select proper safer solvents and antisolvents. The acetone-water with a volumetric ratio of 3:1 was determined to be the most effective safer solvents and antisolvents for extracting 81 wt.% of ABS at room temperature. The recovered ABS has comparable mechanical behavior to the feedstock material with a similar glass transition temperature and molecular weight distribution. Compared to producing virgin ABS material, the dissolution-precipitation process reduces embodied energy by 58%.

154r Reactive Compatibilization of Recycled Polyethylene Terephthalate/Recycled Rubber Particles Blends

Aboulfazl Barati, Erfan Dashtimoghadam

A blend of incompatible polymers such as recycled polyethylene terephthalate (rPET) and recycled rubber particles (RRPs) can have poor mechanical and impact properties because polymeric phases interact weakly and segregate when mixed. This study investigates compatibilization of *r*PET and RRPs blends. Polystyrene-*block*-poly(ethylene-*ran*-butylene)-*block*-polystyrene-*g*-maleic anhydride (SEBS-*g*-MA), poly (ethylene-*co*-glycidyl methacrylate) (EGMA), and poly (ethylene-*co*-methyl acrylate-*co*-glycidyl methacrylate) (EMAGMA) were used as reactive compatibilizers. The rheological properties of the compatibilized blends were characterized and compared with non-compatibilized blends. The results showed that the reactive compatibilizers improved the interfacial adhesion RRPs in *r*PET matrix, leading to enhanced impact strength. Complex viscosity of *r*PET/RRPs blends were found to increase with increasing compatibilizer content, indicating enhanced melt strength and processability. Morphological analysis by scanning electron microscopy revealed a uniform dispersion of RRPs in the compatibilized blends, confirming increased interfacial interactions. These findings suggest that the reactive compatibilization is an effective strategy to optimize *r*PET/RRPs compounds, suitable for continuous manufacturing and melt extrusion.

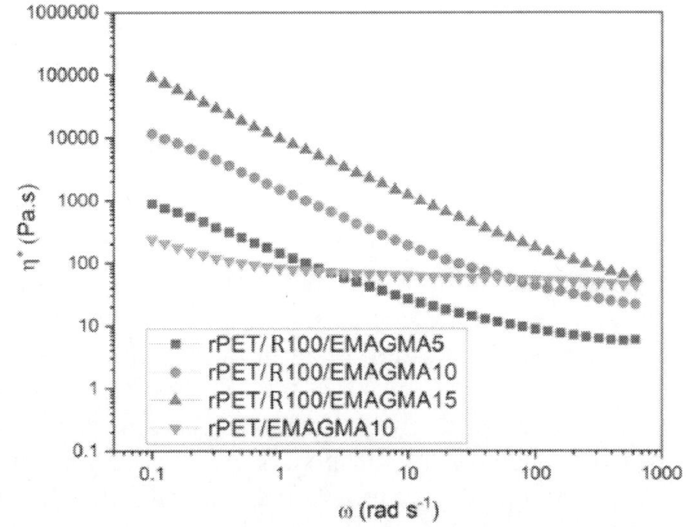

Figure 1. Complex viscosity versus sweep frequency for reactive compatibilized of rPET and rPET/RRP blends

Figure 2. SEM micrographs of cryogenic fractured surface of: (a) R100 and (b) *r*PET/R100/EMAGMA

154s A Two-Stage Process and a Generalized Molecular-Level Kinetic Model for Polyolefin Pyrolysis: Low Molecular Weight Product Evolution

Zhe Fu

Industrialization and development of the global economy have boosted the production of plastics, leading to an explosive growth in the generation of plastic waste. As an undesired result, 6300 million metric tons of plastic waste has been generated globally between 1950 and 2015, 60% of which is polyolefin including polyethylene (PE), polypropylene (PP) and polystyrene (PS), which is one of the most urgent and difficult environmental challenges. In this work, the pyrolysis of PE and PP has been studied by a two-stage tube reactor to obtain high yield of light olefins. The maximum yield of light olefins of PE and PP pyrolysis are 76.1 wt% and 71.8 wt% at 800 °C, respectively. Comparing with the results of other literatures, the present study obtaines a high yield of light olefins and a lower yield of methane. In addition, probable mechanisms for the pyrolysis of PE and PP are proposed to reveal the interplay of reactions which can guide how to produce more light olefins and provide additional information regarding the molecular-level kinetic model. A simplified and generalized molecular-level kinetic model for describing the pyrolysis of polyolefin such as PE, PP and PS has been presented, avoiding the introduction of complex mathematical methods and extensive chemical databases to keep smooth calculation. The structure-oriented lumping (SOL) and Poisson distribution sre selected to transform the feedstock information into the mathematical matrix. The complex reaction network is constructed through automated generation, realizing quick analysis of the data and the delivery of the model parameters. For example, the automatic generated network for PS pyrolysis containing 3,008 substances and 15,025 reactions is partially visualized. Genetic algorithm is used to optimize the kinetic parameters. The model can be applied to predict the experimental data of various researches in the literature without adjusting the rules. Therefore, this work can provide guidance of the recycling and upcycling of plastic waste by analyzing key indicators and predict the impact due to changes in operating conditions from a viewpoint at molecular level.

154t Novel Chemi-Mechanical Recycling Process for Blending of Polyethylene and Polypropylene

Madison R. Reed, Arsalaan Nisar Pathan, Abhishek Banerjee, James M. Eagan, Paul Boulier, Michael T. Timko

Valorizing mixtures of plastics is limited by the cost of separation and improved technologies that produce high-value products form plastic mixtures are desperately needed. Hydrothermal and supercritical water conditions are well known for their use in deconstructing plastics into fuels and other useful products. For example, hydrothermal depolymerization in the absence of oxygen can reduce char formation relative to that obtained under pyrolysis conditions, thereby maximizing yields of more valuable oil products. Addition of oxygen to hydrothermal water results in supercritical water oxidation, a technology that depolymerizes plastics, converts them into CO_2, and releases energy. Accordingly, hydrothermal and supercritical water is often considered exclusively as a deconstruction solvent, with corresponding production of fuels or energy with much less value than the parent polymers. That stated, at temperatures greater than polymer melting and less than rapid depolymerization, hydrothermal water possesses properties that may be appropriate for polymer swelling and moderate chemical rearrangement, including branch and cross link formation. Swelling, branch formation, and cross-linking under hydrothermal conditions therefore has potential as a new method of polymer recycling that incorporates both mechanical and chemical principles and that results in a product with comparable value or even greater value than the plastic feed.

We hypothesize that hydrothermal water can be used for blending immiscible commercial polymers at reaction conditions above polymer melting temperatures but less than depolymerization. The proposed process can be termed "chemi-mechanical" recycling as it aims to create a high-quality blend of ordinarily immiscible polymers, such as polyethylene (PE) and polypropylene (PP). Chemi-mechanical recycling employs a high-temperature reactor and water to blend otherwise immiscible polymers; rapid cooling can preserve the polymers in this blended state, resulting in a polymer-polymer composite with superior properties to those obtained from thermal mixing alone.

Experimental results with PE-PP mixtures indicate that hydrothermal treatment can be categorized into three regimes: 1) a melting regime, under which neither depolymerization nor blending occur, 2) a depolymerization regime, under which waxes or oils can be produced, and 3) an intermediate regime under which compatibilization occurs. In this regime, controlled reactions and solvent swelling promote micro-scale blending resulting in a composite that may have superior properties to those possible from traditional recycling of PE and PP. In this intermediate regime, we observe minimal molecular weight degradation and improved molecular mixing On the other hand, temperatures at the supercritical point of water correspond to the depolymerization regime, producing a brittle material due to partial depolymerization. Temperatures greater than the critical point produce waxes. Results from differential scanning calorimetry (DSC), seen below, show that at elevated temperatures there is a significant loss in both melting temperature (T_m) and percent crystallinity (X_c). However, treatment at temperatures below the supercritical point (SC), results in PE-PP blends which exhibit minimal degradation, as indicated by measured melt temperatures. In addition to melting temperatures, gel-permeation chromatography (GPC) results show that treatment above the critical temperature results in significant loss in molecular weight. At temperatures much less than the critical point, the molecular weight is unchanged after treatment. However, when treated at temperatures near the critical point, only a modest reduction of molecular weight is observed.

As the main issue with these blends is incompatible mixing, SEM was conducted to analyze the phases and degree of mixing. Figure 2a shows SEM images of melt blended PE and PP in the absence of hydrothermal conditions. As seen, the two polymers are not well mixed in this sample and suffer from interfacial adhesion failure. The morphology of the sample treated at a temperature above the critical point can be seen in Figure 2b, which shows intimate mixing of PE and PP. The significant loss in molecular weight observed for this sample means these reaction conditions not ideal for a reusable product. Finally Figure 2c shows well-distributed PP beads in a PE matrix obtained after processing at temperatures less than the critical point, indicating intimate mixing. When combined with Figure 1, we therefore observe both intimate mixing and retention of molecular weight in PE-PP samples treated hydrothermally at temperature below the critical point.

In summary, treatment of a PP-PE mixture in water near its critical point produces an intimate blend which retains its original molecular weight with negligible loss and with retention of its original melting characteristics. The results presented here establish a new, previously undiscovered regime in hydrothermal water that combines chemical and mechanical recycling as a new chemi-mechanical approach that transforms mixed plastic streams into a compatibilized product.

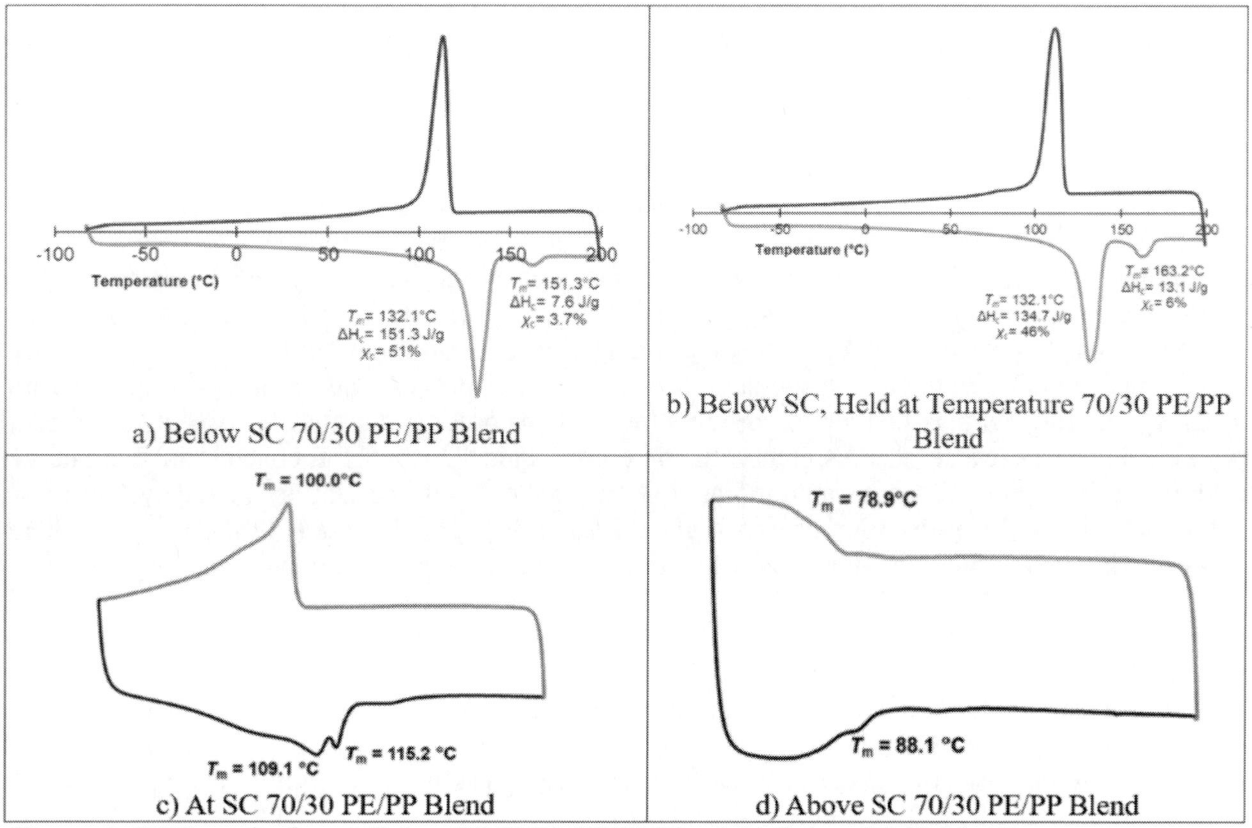

Figure 1: DSC results for: **a)** Below SC 70/30 PE/PP Blend, **b)** Below SC, Held at Temperature 70/30 PE/PP Blend, **c)** At SC 70/30 PE/PP Blend and **d)** Above SC 70/30 PE/PP Blend

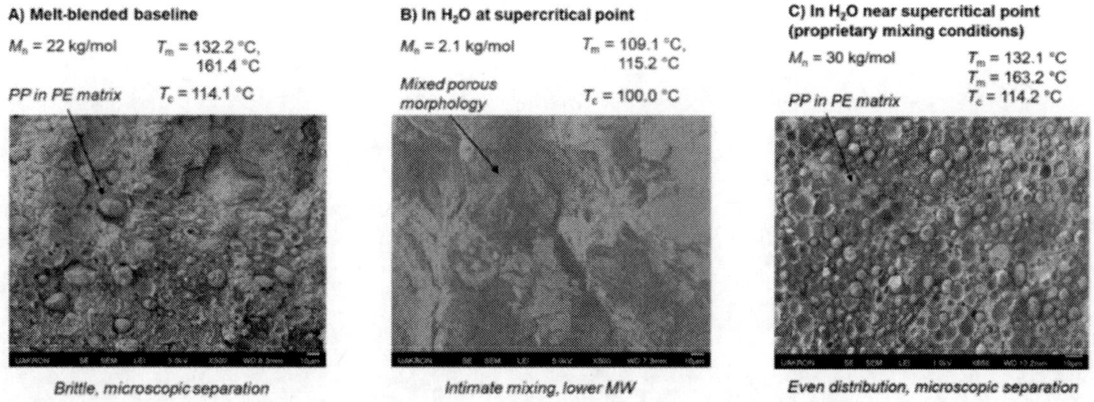

Figure 2: SEM images of: **a)** Melt-blended baseline, **b)** Blend at SC, **c)** Blend below SC held at temperature

154u Sovothermal Upcycling of E-Waste Plastics into Blendstock Fuels and Chemicals

Vahab Ghalandari, Toufiq Reza

Electrical waste plastic (e-waste plastic) refers to the plastic components and housings that are discarded from electronic equipment such as computers, TVs, and mobile phones. Because of the hazardous compounds and heavy metals in electronic components that may leach into the environment and affect human and animal life, e-waste plastic presents a serious threat to the environment and human health. The rising demand for electronic devices and the quick rate of technological progress have made e-waste plastic a significant problem, underlining the necessity of efficient waste management and sustainable upcycling waste process. Solvothermal liquefaction (STL) is a thermochemical conversion process that utilizes polar and nonpolar solvents to turn waste polymers, such as e-waste plastics, into valuable compounds. This work focused to investigate the possibility of utilizing toluene as a nonpolar solvent for the depolymerization of four types of e-waste plastic including polyamide (PA), polycarbonate (PC), polyoxymethylene (POM), and polyether ether ketone (PEEK) into crude oils during STL. Depolymerization experiments were performed using a 7 ml batch reactor at 325-375 °C, and 1-6 h of reaction times. The results indicated that the high temperature and long residence time can enhance the depolymerization of the e-waste plastics. The highest STL conversation was observed for POM at 375 °C and 6 h. Furthermore, the ultimate analysis showed that the crude oil produced from STL of e-waste plastics has an appropriate higher heat value (HHV). Finally, the thermogravimetric analysis (TGA) indicated that most portion of the crude oils is in the range of lighter hydrocarbons. Finally, gas chromatography mass spectroscopy reveals that valuable fuels and chemicals can be formed from e-wastes by STL.

154v Tidytron: Reducing Lab Waste Using Validated Wash-and-Reuse Protocols for Common Plasticware in Opentrons OT-2 Lab Robots

John Bryant Jr., Clay Wright

Every year biomolecular laboratories generate a combined total of ~5.5 million tons of plastic waste in their pursuit of scientific discovery and engineering goals. To increase sustainability and simultaneously reduce recurring costs of biomolecular research, the automation of plastic-ware cleaning using liquid handling robots is a promising solution. However, open-source, validated protocols using low-cost lab robotics for effectively removing contaminants from plasticware—facilitating their safe reuse—have not yet been developed. Here we have developed and validated protocols for cleaning micropipette tips and microtiter plates that are contaminated with DNA, *E. coli*, and *S. cerevisae*. We tested a variety of cleaning solutions, contact times, agitation methods, and useful durations with the aim of minimizing corrosion and cost, while maximizing cleaning stringency and sustainability. We tested and validated these cleaning procedures by comparing fresh versus cleaned tips for contamination with DNA/cells or the cleaning solutions. We assessed contamination by measuring colony forming units by plating, event counts and debris by flow cytometry, and PCR efficiency and DNA concentration by qPCR. Open source cleaning protocols will be made available at https://github.com/PlantSynBioLab/opentrons-sustainability.

154w A Sustainable Hybrid Catalyst to Depolymerize Recycled Polyethylene Terephthalate

Aboulfazl Barati, Somayeh Mohammadi, Mojtaba Enayati, Erfan Dashtimoghadam

Chemical recycling technologies use methods to break down ever-growing plastic waste into valuable materials (*e.g.*, petrochemical feedstock, monomers and building blocks for new plastics, and fuels). Here we propose a green hybrid catalyst for the depolymerization of polyester plastics waste. A catalyst composed of carbon nanodots doped with zinc (Zn-CNDs) on a functional zeolite framework (Zn-CNDs@zeolite) was successfully synthesized. CNDs were synthesized through a one-pot solvothermal method employing pyrolysis of plastic waste. The resulting CNDs were characterized by means of X-ray diffraction, microscopy, UV-Vis, and Raman spectroscopy techniques. The catalytic activity of the Zn-CNDs@zeolite for the depolymerization of recycled polyethylene terephthalate (*r*PET) was evaluated *via* glycolysis in a high-pressure reactor. PET waste from drinking water bottles, ethylene glycol (EG), and the Zn-CNDs@zeolite were feed to a high-pressure reactor and glycolysis of PET was performed at 200°C for 1h at pressures around 2 bar. Our results show that the Zn-CNDs@zeolite can efficiently depolymerize PET to its monomeric unit, bis(2-hydroxyethyl terephthalate (BHET) (or oligomers) by controlling the catalyst loading.

154x Production of Polystyrene Microparticles and Microfibers from Waste Expanded Polystyrene Using Spinning Disk Technique

Kiruthika Jayaseelan, Sreejith C, Shijoy Thomas, Raghunathan Rengaswamy, Basavaraj M. Gurappa

Expanded Polystyrene (EPS) is a non-toxic inert polymer which is used for versatile applications such as materials for packing, decoration purposes, manufacture of cups and trays and so on, due to its fascinating features of low cost, light weight, corrosion-resistance, and processability. Annually, it is estimated that 14.7 MMT of polystyrene is produced, among this 6.6 MMT is EPS. It is found that 6.63 tons/day of polystyrene waste in the form of waste expanded polystyrene (WPS) is produced around the world. WPS causes several environmental problems due to its non-biodegradability and demands considerable space due to its large volume to weight ratio (Chaukura, 2016). Hence disposing or recycling WPS is a challenge. Most of the WPS is disposed of by landfill or by incineration. Landfill causes soil and water pollution and also increases disposal sites. Incineration is an energy recycling method, but incineration of polystyrene produces hazardous chemicals. There are several mechanical recycling methods, but these are more expensive than producing virgin polystyrene. Moreover, these methods are mainly aimed at eliminating polystyrene rather than converting it into a valuable product. Hence, recycling of WPS into a valuable product in an environmentally friendly as well as economical manner is highly necessary (Mangalara, 2016).

Conversion of WPS into valuable products like nano or microparticles and nano or microfibers is a possible way of recycling WPS as they can be used for several purposes like biomedical applications, imaging, preparation of engineered colloid particles, oil adsorption and so on. Particularly, polystyrene microparticles can be used for lateral flow tests, latex agglutination tests, flow cytometry, fluorescence microscopy, and as calibration particles. Polystyrene microfibers with nano-porous structure are considered to be one of the best oil sorbents to absorb oil from the ocean in the event of an oil spill. Polystyrene fibers are also used as fiber membranes and as additives in the paint industry (Mangalara, 2016).

Polymer microparticles with narrow size distribution are preferred due to uniformity in properties, controllable behavior for specific applications and simplified processing. Several existing methods for bulk production of microparticles such as spray drying, fluidized bed drying, electro-spraying and polymerization have limitations especially in the control over particle size distribution and surface morphology. Also, these methods require challenging operating conditions such as high gas flow rates, high temperature for drying, pressures greater than atmospheric conditions and so on. Microfluidic generation of droplets is one method by which highly monodisperse particles can be produced but there are a number of drawbacks concerning lower production rates, handling viscous and impure solutions, separation of particles from continuous phase, wetting issues and so on. Thus, a technique that can overcome all these aforementioned disadvantages and can generate droplets with narrow size distribution is required.

Solvent-antisolvent precipitation can be one method for producing solid particles from microdroplets. In this method, particles are formed when microdroplets of a solution that contains a solvent and a solute come in contact with another solvent (antisolvent) which is miscible with the solvent but does not dissolve the solute (Joye, 2013). Microdroplets can be generated using several atomization techniques such as pressure atomization, air atomization and spinning disk atomization. Air atomization and pressure atomization techniques have similar drawbacks of microfluidic techniques mentioned earlier, as well as produce microdroplets with a broad size distribution. Spinning disk atomization exhibits certain advantages over the other atomization techniques. These include separation of droplets according to their size, no requirement of narrow nozzles, ease of operation and available control variables to control the droplet size from the disk. The droplets are generated when a liquid is introduced to the surface of a high speed rotating disk requiring only moderate pressure to transport the fluids (Ahmed, 2012). Hence, compared to any other atomization techniques, spinning disk atomization when combined with a suitable system for collecting the particles at regions located at different distances from the center of the disk, is found to be an efficient and simple method for producing droplets with a given average size and a narrow size distribution.

In this study, solvent-antisolvent precipitation method is combined with spinning disk atomization (SDA) technique for producing polystyrene microparticles with a narrow size distribution using custom-made SDA equipment. Here, tetrahydrofuran (THF) is chosen as the solvent because of its versatility of dissolving most of the compounds and its miscibility with most of the solvents (Rajeev, 2016). Unlike water which is used as an antisolvent in the production of polystyrene particles by nanoprecipitation, iso-butanol is chosen as an antisolvent. Polystyrene is denser than water, hence when the atomized droplets of polystyrene solution come in contact with water, instead of forming particles; they form circular sheets of solid polystyrene on the surface of water. Hence, iso-butanol, which is less dense than polystyrene, is used as an antisolvent.

In the experiment conducted using the custom-made SDA equipment, the polystyrene-THF solution is atomized using high speed spinning disk to form microdroplets and these droplets are collected in a tray containing iso-butanol located at a desired distance from the center of the disk to form microparticles with a narrow size distribution. The observations from this experiment indicate that the average size of the particles can be tuned by varying: (i) the concentration of EPS, (ii) the speed of the spinning disk, (iii) flow rate of the polystyrene solution, and (iv) location of the antisolvent tray used for precipitation.

In case of polystyrene microfibers production, current techniques like electrospinning and centrifugal spinning have several disadvantages. Electrospinning does not have a higher production rate, requires a high voltage external power source and may not be able to handle all kinds of hydrocarbon solvents and viscous solutions (Liu, 2015). Centrifugal spinning, an emerging method, has several demerits such as lower production rate, handling viscous solutions and clogging issues (Doan, 2019). Hence, an alternative method is required for the production of microfibers that addresses the aforementioned demerits. Spinning disk technique with drying can be used for the production of polystyrene microfibers. In this technique, polystyrene solution at a high concentration fed to the spinning disk with simultaneous gas drying forms polystyrene microfibers.

For the production of microfibers, the concentration of the polystyrene in THF solvent is increased to five times that required for particle production. An additional gas distributor fixed above the spinning disk inside the SDA equipment, supplies nitrogen for drying the fibers. The main parameters for microfiber production are polymer concentration, feed liquid flow rate, speed of the disk and gas flow rate.

There is no method reported to produce both microparticles and microfibers using single equipment. This can be achieved in this custom-made SDA equipment with two separate feed systems where one system contains a highly concentrated polystyrene-THF solution and the other one contains only THF. A static mixer at the downstream of these two feed systems mixes the two liquids before feeding the solution to the disk. By varying the flow rates of the liquids in these two systems and thereby altering the concentration of the feed liquid, microparticles or microfiber production can be achieved.

Analysis of the microscopy images of polystyrene microparticles collected at a particular distance from the spinning disk indicates a narrow size distribution with an average particle size 107 μm. This suggests that, using SDA technique, conversion of EPS to polystyrene microparticles of narrow size distribution can be achieved in a controlled and reproducible manner. The microfibers produced by spinning disk with drying technique have nano-porous structure at the surface with an average fiber size 4 μm. Using SDA equipment with two feed systems and by altering the concentration of the feed liquid, microparticles or microfibers production can be achieved by the same equipment. This equipment and techniques can be used for the conversion of other waste polymers into microparticles and microfibers and can find new applications in various fields such as production of drug particles and microencapsulation of food ingredients.

References:

1. Chaukura, N., Gwenzi, W., Bunhu, T., Ruziwa, D.T. and Pumure, I., 2016. Potential uses and value-added products derived from waste polystyrene in developing countries: A review. *Resources, Conservation and Recycling*, *107*, pp.157-165.
2. Mangalara, S.C.H. and Varughese, S., 2016. Green recycling approach to obtain nano-and microparticles from expanded polystyrene waste. *ACS Sustainable Chemistry & Engineering*, *4*(11), pp.6095-6100.
3. Joye, I.J. and McClements, D.J., 2013. Production of nanoparticles by antisolvent precipitation for use in food systems. *Trends in Food Science & Technology*, *34*(2), pp.109-123.
4. Ahmed, M. and Youssef, M.S., 2012. Characteristics of mean droplet size produced by spinning disk atomizers. *Journal of Fluids Engineering*, *134*(7).
5. Rajeev, A., Erapalapati, V., Madhavan, N. and Basavaraj, M.G., 2016. Conversion of expanded polystyrene waste to nanoparticles via nanoprecipitation. *Journal of Applied Polymer Science*, *133*(4).
6. Liu, W., Huang, C. and Jin, X., 2015. Electrospinning of grooved polystyrene fibers: effect of solvent systems. *Nanoscale research letters*, *10*(1), pp.1-10.
7. Doan, H.N., Nguyen, D.K., Vo, P.P., Hayashi, K., Kinashi, K., Sakai, W., Tsutsumi, N. and Huynh, D.P., 2019. Facile and scalable fabrication of porous polystyrene fibers for oil removal by centrifugal spinning. *ACS omega*, *4*(14), pp.15992-16000.

154y Formation of PET Oligomers Via Microwave-Assisted Heating

Sean Najmi, Brandon Vance, Esun Selvam, Dylan Huang, Dionisios Vlachos

This work investigates the formation of long-chain oligomers (LCO) as an alternative to the full glycolysis of PET. Heating for the reaction is done using microwave irradiation by taking advantage of the excellent microwave susceptor properties of the solvent, ethylene glycol. This enables rapid and selective heating of the solution and the potential for electrification through renewable energy sources. Mn oxide catalysts for depolymerization are investigated because Mn is a cheap and earth-abundant metal. Formation and characterization of the LCO are subsequently investigated. The molecular weights of the LCO change slightly based on the amount of solvent used but are much lower than the starting PET. As reaction time increases, the transesterification reaction proceeds to reduce the molecular weight and dispersion index of the oligomers. High surface area catalyst fully converted the PET to BHET (monomer) at 20 min at 220 °C. Oligomers had molecular weights of around 2000-3000 Da as determined by MALDI-TOF. Their flexibility was analyzed using NMR relaxometry and found to be comparable to PET. Thermal analysis demonstrated that the oligomers possess similar properties to PET. We discuss mechanistic insights into the formation of oligomers and monomers.

154aa Thermal Oxo-Degradation of Post-Consumer Waste Plastics to Increased Yields of Desirable Products Using Less Energy

Jessica Brown, Sarah Tyree, Ryan Smith, Robert Brown, Tannon Daugaard

Lack of attention to the end-of-life disposal of plastics has resulted in their relentless accumulation in the environment or in landfills, motivating research on upcycling, largely defined as depolymerizing plastics into chemical building blocks for new products. One such method is pyrolysis of plastics to commodity fuels and chemicals. However, commercialization of conventional pyrolysis of plastic in an inert atmosphere has been hampered by large thermal requirements stemming from long reaction rates. The rate of thermal depolymerization of waste plastics can be accelerated by the addition of oxygen in a process known as thermal oxo-degradation (TOD). This study offers the prospect of TOD to upcycle post-consumer waste plastics rapidly and efficiently. Thermal oxo-degradation proceeds through a multi-step free-radical chain reaction in which oxygen has a catalytic effect on chain breaking reactions. Using moderate temperatures and small amounts of air in a fluidized bed reactor, we have shown that high density polyethylene and polypropylene are rapidly deconstructed to more than 80 wt.% condensable products while using up to 84% less energy due to exothermic energy release from partial oxidation reactions. Compositional analysis of TOD product oil shows 61 wt.% olefins and 19 wt.% di-olefins, which can be upgraded to value-added products. Similarly, polystyrene can be depolymerized through TOD to high yield (69 wt.%) of the monomer styrene, comparable to the 65 wt.% styrene achieved during pyrolysis. Through TOD, the depolymerization of polystyrene can become a self-sustained reaction, eliminating the need for external provisioning of thermal energy to the process. Comparable yields of the monomer styrene have been produced through this process. Thermal oxo-degradation has proved to be a successful pathway to a plastics circular economy through the reduction of energy while maintaining desirable yields, making it a commercially viable process.

154ab Real-Time Plastic Waste Recognition with Mid-IR Standoff Detection and Advanced Machine Learning

Yaoli Zhao, Patatri Chakraborty, Zixia Meng, Thomas Thundat

Plastic waste identification is vital for recycling and mitigating the effects of global warming. Since plastics do not degrade easily with time, the discarded plastics pose a grave environmental problem. Only less than 10% of plastics is recycled in the US. Also, sorting plastic waste in MRF (Materials recovery facility) is often complicated by the presence of additives and contaminants. At present, sorting of plastic waste is carried out manually which is time consuming, costly and labor intensive. Therefore, developing a sorting technique with high selectivity, sensitivity, and minimal false identification is crucial for increasing plastic recycling rates. We have been developing a sensor platform capable of improving the accuracy of automated sorting of various plastics by using molecular identification using mid-infrared (MIR) spectroscopy. This technique of standoff photothermal detection utilizes mid-infrared photo-thermal spectroscopy for characterizing plastic samples. Since MIR is free of overtones, this region is known as the molecular fingerprint region. In this technique, the plastic surface is illuminated with a pulsed laser beam from a tunable quantum cascade laser (QCL). The photothermal detection is accomplished by collecting the scattered light from the plastic surface which contains the molecular vibration spectrum of the sample as well as the contaminants. A silicon bi-material microcantilever, fabricated by depositing a thin metal film on one of its sides is used as a sensitive uncooled infrared detector. Monitoring the bi-material cantilever bending as a function of illumination wavelength provides the IR absorption spectrum of the plastic material. Our method accurately classifies MIR spectra of plastics with a 100% success rate, even in the presence of additives and contaminant residues, using advanced machine learning techniques. This promising approach enables the rapid and efficient characterization of plastic waste in industrial applications where high throughput is essential. It will help in enhancing the quality of plastic sorting process at materials recovery facilities and other industries by making the process inexpensive and time efficient.

154ae Recovery of Virgin-like Polypropylene from Mixed Plastic Wastes

Parikshit Sarda, Sridhar Viamajala, Joseph Lawrence

Polypropylene (PP) is the most widely produced and used polymer in the world but has an *exceptionally poor recycling rate* of only about 1%. PP and other polyolefins (*i.e.*, various grades of polyethylene (PE)) are often retrieved together as mixture in a "float" fraction during commercial plastic waste recycling. The recovery of high purity PP from these (and other) mixed plastic streams, is the *primary bottleneck* that has prevented the recycle and circular economy of PP. In this paper, we present a sustainable approach to selectively dissolve and recover PP from mixed plastic wastes using waste cooking oils or other triglycerides. These "green solvents" are biodegradable, non-toxic, and bio-sourced. Of all the major polymers tested in our study, *PP alone was soluble* in these solvents while other plastics, including PE, remain insoluble. This exceptional attribute of the triglyceride solvents facilitates the *selective recovery of virgin-like PP from virtually any mixed plastic waste.* The dissolved PP can subsequently be easily re-precipitated with an anti-solvent. An additional advantage of this unique PP-selective dissolution-precipitation approach is that fillers and pigments, commonly present in commercial PP, can also be separated from the polymer. Finally, the triglyceride solvents can be recovered and reused multiple times without loss in performance.

We have *demonstrated high yield (>95%) recovery of high purity, pigment/filler-free PP* from a variety of feedstocks, including floats from commercial automotive waste, post-consumer polyolefin waste (i.e. takeout food containers), commercial PET recycling, discarded surgical masks, and multilayer flexible food packaging (Lays chips). These results are shown in the graphic below. Polymer properties of the recovered PP from these various feedstocks will be presented to demonstrate the virgin-like quality of our product.

Examples demonstrating the recovery of virgin-like PP from mixed plastic waste feed stocks

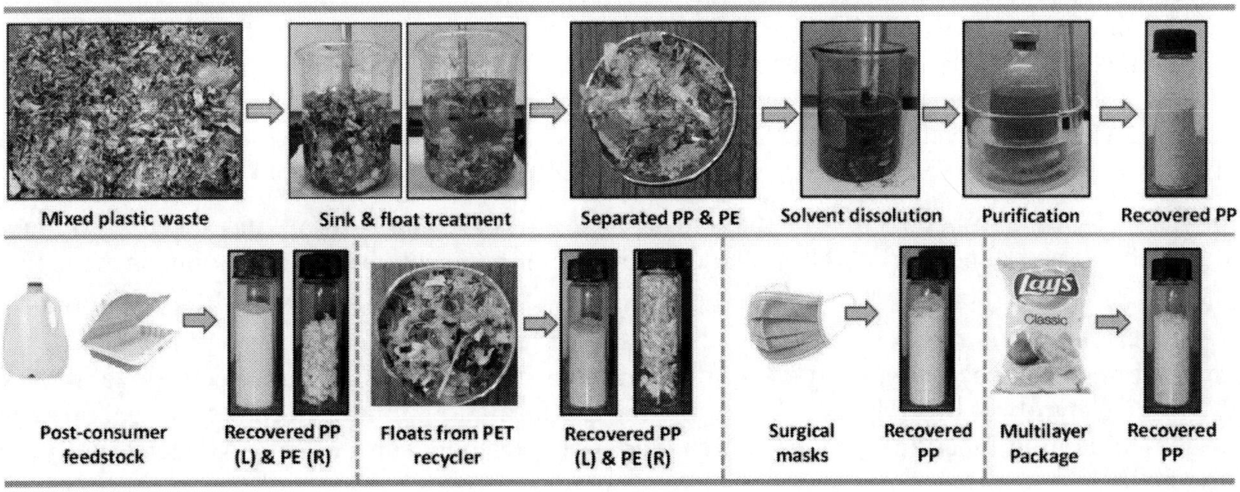

154ah Insights into Co-Pyrolysis of Polyethylene Terephthalate and Polyamide 6 Mixture through Experiments, Kinetic Modeling and Machine Learning

Varaha Jonnalagedda

The global production of plastics increased from 2 million metric tonnes (MMT) in 1950 to 380 MMT in 2015 [1], leading to a subsequent rise in plastic waste [2]. It has been estimated that only 9% of plastic waste produced globally has been recycled [1]. The major cause of the increase in plastic waste is due to the use of durable goods, containers, and packaging [3]. Among several other polymers, PET and polyamide 6 (PA6) are used in durable goods and packaging due to their mechanical strength, mechanical stabilization, and moisture barrier [4,5]. The durable goods like carpets and rugs contain 15 and 25 wt.% of PET and PA6, respectively [6]. The packaging industry is one of the main consumer of plastics and has a share of 36%, followed by construction (16%) and textiles (15%) [7]. The recycling of multilayer packaging is one of the most challenging issues in the solid waste management sector. This is primarily because multilayer packaging is made up of different polymers and materials like metal oxide coatings (AlO_x or SiO_x) and aluminum foil that are difficult to separate and recycle [8,9]. Pyrolysis is a promising thermochemical recycling technology that can accept single streams as well as a mixed stream of plastics to recycle in a single step without the need for separation unlike mechanical recycling [10]. It involves the thermal decomposition of the polymeric structure in an inert ambiance at moderate temperatures of 400-600 °C [11].

In this study, non-isothermal pyrolysis of polyethylene terephthalate (PET), polyamide 6 (PA6), and their mixtures was studied in a thermogravimetric analyzer at different heating rates. The temperature of maximum decomposition (T_{max}) decreased by 25 – 45 °C and 35-55 °C for the PET:PA6 mixtures (3:1, 1:1, 1:3) compared to PET and PA6, respectively. The kinetic analysis was initially carried out using isoconversional method. However, the dependency of activation energy on conversion was observed for the co-pyrolysis of PET and PA6 that suggested the occurrence of multi-step reactions in the mixtures. Distributed activation energy model (DAEM) was used in this study to describe the multistep reactions occurring during pyrolysis of PET:PA6 mixtures. In this work, a four-parallel reaction DAEM was developed to describe the pyrolysis kinetics of PET:PA6 mixtures. The mean activation energies (E_o) for PET, PA6, and mixtures varied in the range of 244 – 255, 140 – 215, and 140 – 255 kJ mol^{-1}, respectively. The mass loss profiles of PET and PA6 mixtures were also modeled using artificial neural network (ANN). Out of 155 ANN models, the best prediction was made by ANN511 with $R^2 > 0.997$ for both test and unseen data. The interaction effects observed through TGA experiments and subsequent kinetic analysis were further assessed in terms of product composition using analytical pyrolysis coupled with gas chromatograph/mass spectrometer (Py-GC/MS). Co-pyrolysis of PET and PA6 resulted in the formation of new aromatic compounds with nitrogen-containing functional groups, which were not detected when PET or PA6 were pyrolyzed individually.

References

[1] R. Geyer, J.R. Jambeck, K.L. Law, Production, use, and fate of all plastics ever made, Science Advances. 3 (n.d.) e1700782. https://doi.org/10.1126/sciadv.1700782.

[2] S.M. Al-Salem, P. Lettieri, J. Baeyens, Recycling and recovery routes of plastic solid waste (PSW): A review, Waste Management. 29 (2009) 2625–2643. https://doi.org/10.1016/j.wasman.2009.06.004.

[3] O. US EPA, National Overview: Facts and Figures on Materials, Wastes and Recycling, (2017). https://www.epa.gov/facts-and-figures-about-materials-waste-and-recycling/national-overview-facts-and-figures-materials (accessed March 30, 2022).

[4] K. Kaiser, M. Schmid, M. Schlummer, Recycling of Polymer-Based Multilayer Packaging: A Review, Recycling. 3 (2018) 1. https://doi.org/10.3390/recycling3010001.

[5] C.T. de M. Soares, M. Ek, E. Östmark, M. Gällstedt, S. Karlsson, Recycling of multi-material multilayer plastic packaging: Current trends and future scenarios, Resources, Conservation and Recycling. 176 (2022) 105905. https://doi.org/10.1016/j.resconrec.2021.105905.

[6] I. Sas, J.A. Joines, K.A. Thoney, R.E. King, Logistics of carpet recycling in the U.S.: designing the collection network, The Journal of The Textile Institute. 110 (2019) 328–337. https://doi.org/10.1080/00405000.2018.1480101.

[7] International Energy Agency, The Future of Petrochemicals, (2018) 132.

[8] T.W. Walker, N. Frelka, Z. Shen, A.K. Chew, J. Banick, S. Grey, M.S. Kim, J.A. Dumesic, R.C. Van Lehn, G.W. Huber, Recycling of multilayer plastic packaging materials by solvent-targeted recovery and precipitation, Sci Adv. 6 (2020) eaba7599. https://doi.org/10.1126/sciadv.aba7599.

[9] Z. Tartakowski, Recycling of packaging multilayer films: New materials for technical products, Resources, Conservation and Recycling. 55 (2010) 167–170. https://doi.org/10.1016/j.resconrec.2010.09.004.

[10] K. Ragaert, L. Delva, K. Van Geem, Mechanical and chemical recycling of solid plastic waste, Waste Management. 69 (2017) 24–58. https://doi.org/10.1016/j.wasman.2017.07.044.

[11] O. Dogu, M. Pelucchi, R. Van de Vijver, P.H.M. Van Steenberge, D.R. D'hooge, A. Cuoci, M. Mehl, A. Frassoldati, T. Faravelli, K.M. Van Geem, The chemistry of chemical recycling of solid plastic waste via pyrolysis and gasification: State-of-the-art, challenges, and future directions, Progress in Energy and Combustion Science. 84 (2021) 100901. https://doi.org/10.1016/j.pecs.2020.100901.

154aj MD Simulation of Waste Plastic-Asphalt Compatibility

Aniruddha Chowdhury, Andrew J. Peters, Pouria Nourian, Nazimuddin Wasiuddin

In order to improve the longevity of asphalt pavements, which require frequent maintenance owing to rutting and cracking, this study examines the application of polymer modification. Asphalt binders with polymer modifications have demonstrated enhanced performance characteristics, such as increased resistance to rutting, cracking, and oxidative long-term aging. Waste plastics have been looked at as a potential replacement for virgin polymers in asphalt binders, offering economic savings and environmental advantages. However, the compatibility between waste plastics and binder remains problematic, and better methods of screening plastics and binders for improved interaction strength are required. The phase behavior of asphalt-polymer blends is predicted using the Flory-Huggins interaction parameter, and the compatibility of various polymers with asphalt molecules is examined using MD modeling. The analysis of four different polymers reveals that polyethylene is the most compatible and that the phase separation behavior of an asphalt-polymer blend is significantly influenced by the molecular weight of the binder. This study lays the groundwork for future research into the modification of polymers and the use of waste plastic in asphalt binders.

154ak Upcycling Plastics to Value-Added Chemicals Via Electrified Spatiotemporal Heating

Qi Dong, Liangbing Hu

Nearly 9 billion metric tons of plastic have been produced in the human history, among which about 2/3 have become waste that can result in long-term negative environmental and ecological impact. For example, over 14 million tons of plastic waste end up in the ocean every year, threatening biodiversity and wildlife. Moreover, microplastics and its microfibers have been often detected in air and municipal drinking water, causing long-term health concerns. With the continuing heavy demand for plastic products as well as the aforementioned issues, it is urgent to develop efficient plastic recycling strategies to ensure a sustainable future. Among all thermochemical plastic recycling routes, depolymerization is a promising one as it can turn waste plastic into constituent monomers for further polymerization or as chemical feedstocks for manufacturing. However, many commodity plastics cannot be depolymerized with high selectivity and yield, as it is challenging to control the reaction progress and pathway using conventional thermochemical treatment. Although catalysts can be applied to improve selectivity, they are prone to performance degradation, which limits the system durability.

In this talk, I will demonstrate a catalyst-free depolymerization method via the pyrolysis reaction that can convert model polyolefin and polyester plastics to their monomers with high yield (Nature 2023, accepted). This pyrolysis process is realized by two critical features: (1) a spatial temperature gradient and (2) a temporal heating profile. The spatial temperature gradient is enabled by a bilayer structure made of porous carbon felts, in which the top layer is electrified to conduct heat down to the underlying reactor layer and plastic feed. The resulted temperature gradient promotes continuous melting, wicking, vaporization and reaction of the plastic as it experiences increasing temperatures when traveling upward, enabling a high depolymerization degree. In the meantime, we apply pulsed electrical heating through the top heater layer, creating a temporal heating profile that features periodic high temperatures (e.g., ~600 °C) to enable high reaction rate and conversion, yet the transient heating duration (e.g., 0.11 s) can reduce unwanted side products and keep the reaction under far-from-equilibrium conditions. With this process and setup, we converted polypropylene and polyethylene terephthalate to their monomers via the pyrolysis reaction. We achieved monomer yields of ~40%, which are among the highest compared to conventional thermochemical methods even with optimized catalysts. Owing to the high tunability of the electrified heating (temperature, timescale), this approach potentially offers a unique platform to convert plastic waste to a range of value-added chemicals.

154al Upcycling of Waste Poly(vinyl chloride) (PVC) through Depolymerization Under Mild Conditions

Ali Alshaikh, Jason Bara

Interest is growing in managing plastic wastes through depolymerization and upcycling. While certain commodity polymers like poly(ethylene terephthalate) (PET) have received a great deal of attention, poly(vinyl chloride) (PVC) has received much less attention. PVC has very large worldwide production volumes, while simultaneously being among the least recycled plastics. Undoubtedly, PVC has unique challenges associated with recycling, specifically, its propensity to undergo dehydrochlorination and crosslinking during thermomechanical recycling, and release of HCl. Here, we present options to manage and valorize PVC that take advantage of dehydrochlorination and enable depolymerization toward small molecules and oligomers that can be upcycled. Dehydrochlorinated PVC [DHPVC] with conjugated polyene segments was produced in our lab followed by catalytic hydrogenation at ambient temperature and relatively low pressures. This hydrogenation acts as a preparation step, as it greatly lowers (or eliminates) the presence of conjugated C=C bonds in the modified PVC, while simultaneously stabilizing it from oxidation and/or crosslinking. This hydrogenated dehydrochlorinated PVC [H_2-DHPVC] can then be cleaved via olefin metathesis catalysts to yield small molecules and oligomers. This presentation will discuss the benefits and challenges of working with PVC, as well as an overview of methods and target products.

154ao Upcycling Virgin and Waste Polyethylene to Reprocessable Dynamic Covalent Networks Via Free-Radical Grafting of Dialkylamino Disulfide Bonds

Logan Fenimore, Boran Chen, John Torkelson

Plastics enable modern life through their advantageous properties and broad applicability. Regardless of their type or use, plastics are challenging to recycle efficiently. Current methods for recycling spent thermoplastics such as re-extrusion with additives result in property degradation over time and the relegation of these downcycled polymers to low-value applications. Plastics may be permanently cross-linked into thermosets, yet permanent cross-links prevent these plastics from being processed and molded into new shapes at high temperature. An emerging avenue to mitigate these sustainability problems involves enriching waste plastics with dynamic covalent bonds as chemical cross-links. By introducing dynamic covalent cross-links, previously thermoplastic materials exhibit robust mechanical properties characteristic of conventional thermosets yet maintain their reprocessability at high temperatures. Using reactive batch processing, we upcycled virgin and waste low-density and high-density polyethylene into covalent adaptable networks (CANs) via melt-state, free-radical grafting of a cross-linker capable of dynamic dialkylamino disulfide (BiTEMPS) chemistry onto polymer chains. Unlike PE thermosets, our PE CANs are reprocessable and recover their thermomechanical properties after reprocessing. We have further shown that, in the absence of crystallinity, high-temperature creep behavior of the CANs is dominated by the exclusively dissociative reversible dynamic chemistry of the cross-linker. This observation also demonstrates the utility of this dissociative dynamic chemistry of high activation energy at suppressing creep in networks exhibiting different viscoelastic behavior. We will also describe cases of other thermoplastics that are able to be upcycled into CANs using our simple method.

154ap Mechanochemical Degradation of Polyethylene into Waxy Residue, Olefin Containing Products and Gaseous C1-C6 Products

Laura Wilcox, Douglas P. Theberge, Yagnaseni Ghosh

Plastics have become widespread in their use with the majority of plastic being single use, thus, leading to approximately 300 million tons of plastic waste are generated annually. The recycling industry and infrastructure, however, is not equipped to handle this magnitude of plastic leading to only ~10-15% being recycled and the remaining waste being landfilled, incinerated, or polluted to the environment (i.e., oceans).[1, 2] In order to combat the plastic waste crisis, chemical recycling to convert waste into value-added products, such as hydrocarbon fuels and petrochemicals has been explored. Current methods include pyrolysis to convert plastic waste into hydrocarbon fuels by using high temperatures (500-1000 °C) in the absence of oxygen to thermally crack the polymer chains into lower molecular weight fractions.[3] The formation of these lower MW is implanted in refinery processes to convert to chemicals as well as may be used as transportation fuel. Although pyrolysis has resulted in production of olefins, the many drawbacks, including catalyst coking, high energy requirements, and a relatively large carbon footprint, limit the implementation of this technology. [4-8] To overcome the drawbacks and challenges inherently present in pyrolysis methods, this work targets sustainable solutions to combat plastic waste using a greener method of conducting chemistry known as mechanochemistry. Mechanochemistry is the process of conducting a solid-state chemical reaction in an environment under extreme mechanical stress, and here, a planetary ball mill is used for these reactions with the major advantages including operation without the use of solvents, reduced reaction times, and changes in product selectivity, and being less energy intensive.[9]

In this work, polyethylene (MW = 4000 da) was degraded using mechanochemical reactions catalyzed with aluminosilicate materials of varying properties such as Si/Al and topology (i.e., FAU and MFI zeolites). Mechanochemical reactions were conducted in an oxygenated environment at ambient temperature for up to 12 h using a planetary ball mill (Retsch PM 400) with stainless-steel jars containing grinding balls, polyethylene, and the aluminosilicate catalyst. Post reaction, a series of steps were taken to extract the formed olefin-containing residue from the catalyst and stainless-steel grinding balls including sonication, centrifugation, and rotary evaporating. Following the extraction of the residue, thermogravimetric analysis (TGA) was measured on the resulting product and revealed the product degraded faster and at a lower temperature than the reference polyethylene. ^{1}H, ^{13}C, and Heteronuclear Single Quantum Coherence (HSQC) Nuclear Magnetic Resonance (NMR) spectra were collected and revealed the presence of olefinic peaks in the range of ~5-5.3 ppm (^{1}H) and ~170-200 pm (^{13}C) with HSCQ allowing for a correlation between proton and carbon NMR spectra to determine the bond correlations. Additionally, gaseous products were extracted and injected into a gas chromatography-mass spectrometer (GC-MS) containing a specialized plot column allowing for full separation of the gaseous mixture into its individual components in order to identify each product through both a flame ionization detector and mass spectrometer. Current efforts have been able to both qualitatively and reproducibly identify gaseous products as C1-C6 hydrocarbons with future efforts moving towards the quantification of each species as well as identifying products that form during the reaction as a waxy residue. These exciting results are directing future efforts to optimize the reaction conditions to enhance the formation of valuable chemicals directly from polyolefin plastic feedstock, move towards the quantification of gaseous products, and investigate the recyclability of the catalyst. Overall, this low temperature, energy efficient mechanochemical process demonstrates that polymers can be successfully degraded to lower MW residues containing olefinic functional groups. This reveals the ability and effectiveness of mechanochemistry in degrading polymers into smaller constituent building blocks, including olefins and alkanes, thus paving way for greener and energy efficient solutions to breakdown polymers into monomers.

1. Parker, L. *A Whopping 91 Percent of Plastic Isn't Recycled*. National Geographic Society, 2019.

2. Programme, U.N.E., *Single-use plastics: A roadmap for sustainability*. 2018, United Nations Environment Programme.

3. Jiang, J., et al., *From plastic waste to wealth using chemical recycling: A review*. Journal of Environmental Chemical Engineering, 2022. **10**(1).

4. Li, Q., et al., *Progress in catalytic pyrolysis of municipal solid waste*. Energy Conversion and Management, 2020. **226**.

5. Xue, Y., P. Johnston, and X. Bai, *Effect of catalyst contact mode and gas atmosphere during catalytic pyrolysis of waste plastics*. Energy Conversion and Management, 2017. **142**: p. 441-451.

6. Wang, X., et al., *Soot formation during polyurethane (PU) plastic pyrolysis: The effects of temperature and volatile residence time*. Energy Conversion and Management, 2018. **164**: p. 353-362.

7. Santos, B.P.S., et al., *Petrochemical feedstock from pyrolysis of waste polyethylene and polypropylene using different catalysts*. Fuel, 2018. **215**: p. 515-521.

8. Huang, J., et al., *Chemical recycling of plastic waste for sustainable material management: A prospective review on catalysts and processes*. Renewable and Sustainable Energy Reviews, 2022. **154**.

9. Howard, J.L., Q. Cao, and D.L. Browne, *Mechanochemistry as an emerging tool for molecular synthesis: what can it offer?* Chemical Science, 2018. **9**: p. 3080-3094.

154aq Evaluating the Economic and Environmental Benefits of Deploying a National-Scale, Thermo-Chemical Plastic Upcycling Infrastructure in the United States

Evan Erickson, Philip Tominac, Jiaze Ma, Victor Zavala

Global rates of plastic production, today reaching 380 million tons per year [1], demand comprehensive and reliable plastic waste management systems to avoid the worst impacts of terrestrial and marine plastic pollution [2]. However, overall plastic recycling rates remain well below 10% [1], while the bulk of plastic wastes are incinerated, landfilled, or allowed to accumulate in the biosphere [3].

Traditional plastic recycling by mechanical methods is a mature technology but faces significant inadequacies as a plastic management strategy given that it has been developed only to accept limited types of plastics (mainly polyethylene terephthalate and high-density polyethylene, or PET and HDPE), which generally represent less than 12% of consumer plastic waste [4]. Furthermore, its products are of lower quality than its starting materials [5]. The result of this "downcycling" of plastic materials is an unsustainable system in which inputs can only be recycled a finite number of times before having no remaining use [6]. These factors reinforce the parallel need for environmentally-damaging disposal methods such as landfilling and incineration to accept the waste products not fit for downcycling.

An alternative strategy for plastic waste management, known as plastic "upcycling," uses chemical methods rather than mechanical ones. Upcycling involves a chain of processing technologies: beginning with the sorting and purification of plastic fractions of common recyclable waste streams at municipal recovery facilities (MRFs) and plastic reprocessing facilities (PRFs), plastic wastes undergo a series of thermo-chemical conversions including pyrolysis and steam cracking to arrive at pure plastic monomers, which can finally be repolymerized into products (here, low-density polyethylene and polypropylene, or LDPE and PP) with the same properties as virgin plastics [1]. This capability of forming identical quality products provides a central contrast with traditional mechanical recycling: chemical upcycling is a strategy which approaches a circular plastics economy, while mechanical downcycling does not. Moreover, the thermo-chemical approach provides a flexible avenue to deal with diverse plastic waste types that cannot be mechanically recycled.

In order to accelerate the deployment of plastic upcycling technologies, it is necessary to study its economic viability and its integration with existing waste management infrastructure. Ma et al. [4] recently investigated the implementation of upcycling technologies in the US Midwest region, but no studies have been reported on economic potential at a national scale. Studying economic viability at a national scale is important because it is necessary to determine if there is sufficient plastic waste feedstock available to justify large capital investments and because it is necessary to understand potential regional synergies that can help reduce costs, improve environmental outcomes, and ultimately lead to more circular plastic economies.

Our study uses a large-scale computational framework that integrates techno-economic analysis, value chain analysis, and optimization to examine the potential of deploying an upcycling infrastructure at a national scale. Our results indicate that upcycling provides a vastly more favorable economic picture than current downcycling infrastructure; specifically, we find that deploying an upcycling infrastructure can activate a value chain that generates significant revenue to diverse stakeholders (e.g., chemical industry, waste collectors, and material recovery facilities). Our analysis also finds that deploying such an infrastructure can lead to significant environmental benefits over the common disposal methods of landfilling and incineration.

1. Houqian Li, Horacio A. Aguirre-Villegas, Robert D. Allen, Xianglan Bai, Craig H. Benson, Gregg T. Beckham, Sabrina L. Bradshaw, Jessica L. Brown, Robert C. Brown, Marco Antonio Sanchez Castillo, Victor S. Cecon, Julia B. Curley, Greg W. Curtzwiler, Son Dong, Soumika Gaddameedi, John E. Garcia, Ive Hermans, Min Soo Kim, Jiaze Ma, Lesli O. Mark, Manos Mavrikakis, Olumide O. Olafasakin, Tim A. Osswald, Kostas G. Papanikolaou, Harish Radhakrishnan, Kevin L. S´anchez-Rivera, Khairun N. Tumu, Reid C. Van Lehn, Keith L. Vorst, Mark M. Wright, Jiayang Wu, Victor M. Zavala, Panzheng Zhou, and George W. Huber. Expanding Plastics Recycling Technologies: Chemical Aspects, Technology Status and Challenges. June 2022. DOI: https://doi.org/10.26434/chemrxiv-2022-9wqz0-v2.
2. UNEP. From Pollution to Solution: A global assessment of marine litter and plastic pollution. Technical report, October 2021. Section: publications. URL: http://www.unep.org/resources/pollution-solution-global-assessment-marine-litter-and-plastic-pollution.
3. Qidong Hou, Meinan Zhen, Hengli Qian, Yifan Nie, Xinyu Bai, Tianliang Xia, Mian Laiq Ur Rehman, Qiushi Li, and Meiting Ju. Upcycling and catalytic degradation of plastic wastes. *Cell Reports Physical Science*, 2(8):100514, August 2021. URL: https://www.sciencedirect.com/science/article/pii/S2666386421002186, doi:10.1016/j.xcrp.2021.100514.
4. Jiaze Ma, Philip Tominac, Horacio Aguirre-Villegas, Olumide Olafasakin, Mark Mba-Wright, Craig H. Benson, George W. Huber, and Victor M. Zavala. Economic Evaluation of Infrastructures for Thermochemical Upcycling of Post-Consumer Plastic Waste. October 2022. DOI: https://doi.org/10.26434/chemrxiv-2022-v838h.
5. Coralie Jehanno, Jill W. Alty, Martijn Roosen, Steven De Meester, Andrew P. Dove, Eugene Y.-X. Chen, Frank A. Leibfarth, and Haritz Sardon. Critical advances and future opportunities in upcycling commodity polymers. *Nature*, 603(7903):803–814, March 2022. Number: 7903 Publisher: Nature Publishing Group. DOI: https://doi.org/10.1038/s41586-021-04350-0.
6. Ayah Alassali, Caterina Picuno, Zhi Kai Chong, Jinyang Guo, Roman Maletz, and Kerstin Kuchta. Towards Higher Quality of Recycled Plastics: Limitations from the Material's Perspective. *Sustainability*, 13(23):13266, January 2021. Number: 23 Publisher: Multidisciplinary Digital Publishing Institute. URL: https://www.mdpi.com/2071-1050/13/23/13266, doi:10.3390/su132313266.

154ar Microwave-Assisted Chemical Recycling of Polyolefinic Plastics for the Production of Monomers

Leilei Dai Sr., Roger Ruan Sr.

Current global plastic production is around 380 million tons/year, only 14% of which is recycled. Most of them ends in the landfill, the environment, or the incineration plant, polluting our ecosystem and leading to micro/nano-plastic issue. Developing an effective pathway to remove waste plastics from landfill and incineration plants and create a circular economy requires a more appropriate technology beyond the conventional mechanical recycling by melting and re-molding. Pyrolysis has shown the great potential of plastic recycling for energy, fuels, chemicals, and materials production, enabling the plastic wastes to stay in the economy and out of the environment. However, plastic recycling through catalytic pyrolysis currently suffers from a range of limitations, including: unreliable pyrolysis/catalysis reactor system, short catalyst lifetime, and low economic feasibility. We have proposed to recycle plastic waste based on our catalytic microwave-assisted pyrolysis (CMAP) technology, where the plastic waste is converted into monomers for new plastic manufacturing.

Developing a novel CMAP technology to convert waste plastics to monomers for new plastic production requires highly efficient and reliable system and catalysts. In the past few years, we developed a novel system of CMAP featuring a mixing SiC ball bed for monomer production from polyolefinic plastics with promising economic and environmental outlook. In terms of catalysts, zeolites are considered the most promising cracking catalyst. It has been reported that the increase in acidity and accessibility of acid sites can decrease the degradation temperature and the activation energy, and improve the selectivity of desirable products. However, no clear descriptions for the structure vs performance relationship during catalytic cracking of waste plastics have been developed. Clearly, extending catalyst lifetime without affecting the cracking performance is one of the most important objectives in developing new catalysts for plastic cracking. In our study, we present a comprehensive investigation on the effect of BAS and pore structure on the service lifetime of ZSM-5 zeolites in the plastic cracking. By developing different ZSM-5 zeolites with gradient BAS concentrations, and a range of meso-porosity, we have attempted to establish a correlation between BAS structure and catalyst lifetime, which could be applied to fast catalyst screening in the booming plastic waste pyrolysis industry. Results indicated that the plastic conversion presents significant declining trends with time on stream, indicative of the catalyst deactivation. The experimental data show that both the lifetime and product selectivity of the catalysts changed when their Bronsted acid site (BAS) density and pore structure were altered through the modifications. To quantitatively describe the relationship between them, we developed a BAS surface density (C_B) factor by normalizing the BAS concentration (nBAS) by the BET area (SBET). The relationship between C_B and high density polyethylene (HDPE) throughput is not a linear correlation. Instead, both high and low C_B lead to relatively short catalyst lifetime. Only when BAS densities are located at the middle range (around 0.5-0.9), can zeolite perform the best in terms of lifetime. High BAS densities would mean closer proximity for acid sites, thereby resulting in faster deactivation due to higher coking rates. However, too low BAS densities also result in short lifetime because BAS plays a vital role during cracking reactions.

To improve the catalyst lifetime, a hierarchically ordered macro-meso-microporous nano-sized zeolite catalyst has been developed for catalytic cracking of polyolefinic plastics for monomers (arenes and light olefins) production. The catalyst we developed is able to last for more than 30 h in this reaction at the weight hourly space velocity of 20 /h. Our efforts will advance this technology to a larger commerical operation and help resolve the plastic waste issue facing the society.

239e Sustainable Process Design for Valorization of PET Waste

Sweta Zode

The global plastic waste crisis has become an urgent environmental issue in recent years, with waste polyethylene terephthalate (PET) being a significant contributor. PET plastic is not only non-biodegradable, but it also has low recycling rates reported across the globe. The European Commission estimated in 2017 that only small fraction of PET plastic generated in the EU ends up in recycling. This ultimately leads to landfilling or incineration of the remaining of waste plastic resulting in increased GHG emissions, marine pollution, and human health risks. To tackle the pollution generated from PET plastic waste it is essential to find strategies to prevent PET from reaching landfills. A promising alternative is the use of upcycling technologies for the valorization of PET. This approach holds the promise of reducing the amount of waste by developing circular economies, and its advocates claim that economic benefits may be harnessed by carefully selecting the upcycling reactions. In recent years a significant amount of attention has been devoted to this problem by experimental groups, and different valorization strategies based on pyrolysis, gasification, and other catalytic reactions have been proposed by Khumthai in Energy Conversion and Management journal. However, in the case of PET little attention has been paid to the issue from the perspective of the process and systems engineering community. Most of the work has been done on a case-by-case basis. That is, for selected upcycling strategies some processes have been developed and their economic and environmental impacts analyzed. However, the current state of the art is much larger than a handful of alternatives, and we need to develop tools enabling the accurate and rapid surveying of this design space such that optimal upcycling strategies are identified based on their economic potential and environmental impacts. To overcome this challenge, in this work, we develop a comprehensive superstructure-based optimization framework for PET upcycling. The superstructure is comprehensive because (1) it considers a representative sample of current PET upcycling technologies and (2) it includes a varied set of possible products (monomers, fuels, and specialty chemicals). The problem is formulated as a multi-objective mixed integer nonlinear program to optimize the net present value and unit life cycle environmental impacts. In our work we discuss optimal upcycling strategies, and the trade-off between environmental and economic objectives.

354a Graphitic Carbons By Upcycling Consumer Plastic Waste By Graphene Oxide Additives: Mechanism(s) and Characterization

Akshay Gharpure, Margaret Kowalik, Adri van Duin, Ramakrishnan Rajagopalan, Randall Vander Wal

In the United States, approximately 37 million tons of plastic are used every year according to the McKinsey estimates. Packing and food-service uses contribute 16 million tons which is single use disposable plastic. More than 70% of the single use plastic waste is sent to landfills. This presents a huge opportunity to upcycle the plastic waste into high value carbon materials. Concurrently a big surge in demand for carbon materials is expected from energy storage and transportation applications. Present electric vehicles require ~70 kg graphite. An EV adoption of 1% of the new car market will potentially increase graphite demand by 10%.

When heated, plastics typically crack into light gases through chain unzipping and β-bond scission. This translates into low carbon yield and non-graphitizable residual carbon. We propose an innovative approach of employing GO as templating agent and closed reactor carbonization under autogenic pressure to increase yield and graphitic quality of carbons. Oxygen functional groups on GO would provide the necessary stabilization while the sp2 framework would serve as a template guiding reconstruction of polymer chains into graphitic material. The latter postulated templating effect could realize significant cost and energy savings by enabling graphitic structure at lower temperatures. Notably, a portion of this graphitic material could be used to derive the GO for templated graphitization of fresh feedstock, thus contributing to circular process and upcycling plastic economics. Upcycling overabundant plastic waste into graphitic carbons will reduce pollution, cost and CO2 emissions. Four commercially recycled plastics and their composites with GO have been used to evaluate comparative yield and graphitic quality of the obtained carbon materials. The key advantage of GO as additive is that it nets a substantial increase in carbon yields, nearly 300% in some plastics. GO/plastic composites possess lattice parameters comparable to graphitized anthracene coke.

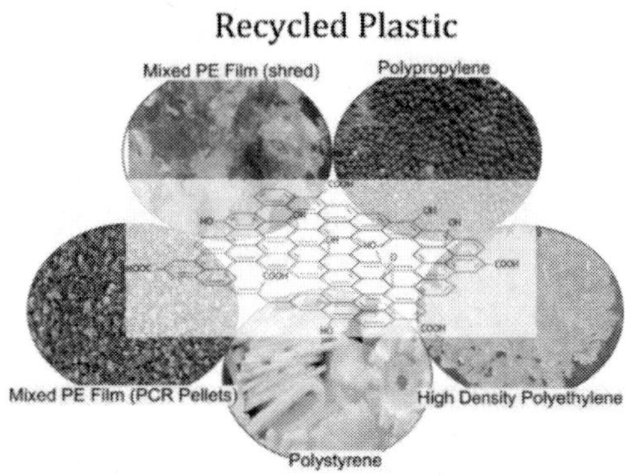

354b Bioconversion of Polyethylene after Thermal Oxo-Degradation As Plastic Upcycling Strategy

Efrain Rodriguez-Ocasio, Mark Blenner, Laura R. Jarboe

Engineered microorganisms can be instrumental in the fight against plastic wastes and for the advancement of a circular economy. Over 8 billion tons of plastics have been produced, and over half of it has ended or will end in landfills or the environment. Plastic accumulates in the environment because evolution has not had enough time to develop a mechanism to break down plastics into molecules that can be recycled by nature. Instead, it can take hundreds of years for plastics to break down via abiotic action, making it the rate-limiting step in plastic biodegradation. Thermal oxo-degradation (TOD) can accelerate this rate-limiting step by rapidly deconstructing the polymer backbone and inserting oxygen functionalities to produce lipids and hydrocarbons. At this point, microorganisms can be employed to convert the carbon that originated from plastic wastes into higher-value products. In this work, non-conventional yeasts were selected against model compounds representative of the composition of thermally oxo-degraded High-Density Polyethylene (HDPE), the most common type of plastic. The down-selected microorganism grew using TOD products from plastics as its sole carbon source without emulsification, overcoming the mass transfer limitations presented by a hydrophobic and solid substrate. The already remarkable growth profile of the microorganism was improved via adaptative laboratory evolution, increasing the growth rate by >200%. This work represents the proof-of-concept for a promising plastic waste upcycling approach and presents a non-conventional yeast with the potential to become a microbial cell factory.

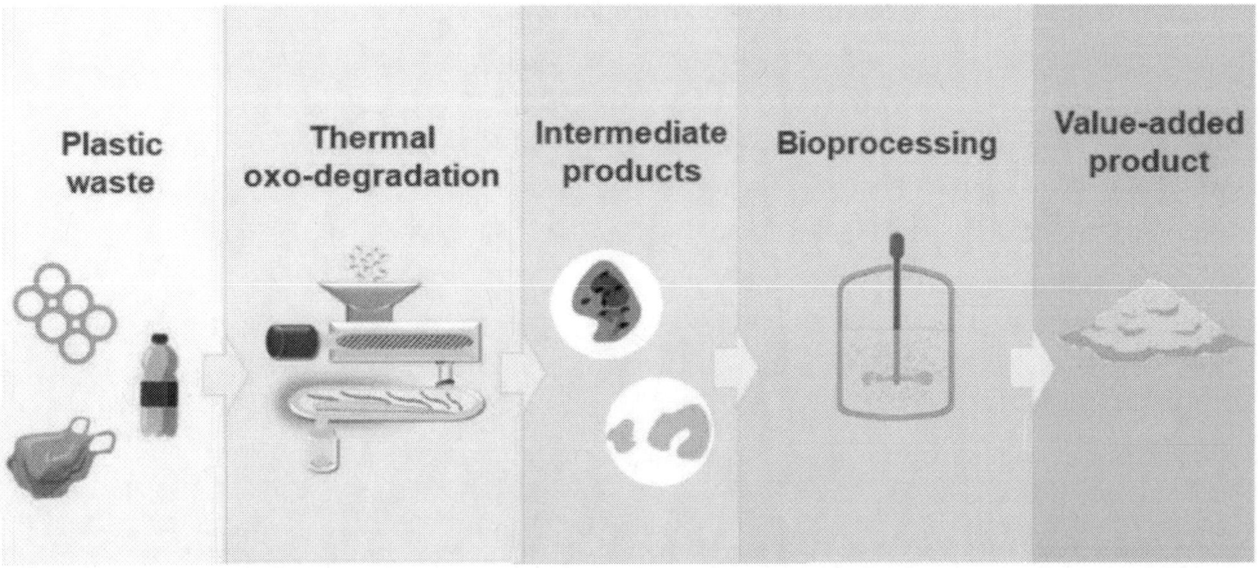

354c Recycling Plastics in the U.S.: Polypropylene Flows and Recycling Processes

Raymond Smith, Sudhakar Takkellapati

Plastic production is currently projected to grow beyond our capabilities to manage. Flows of postconsumer use plastics continue to increase, and so it is important to understand the management pathways for them. A previous study examined PET flows and processes [1]. This effort focuses on polypropylene (PP, Plastic #5), its material flows and recycling processes. The presentation will describe the PP items being collected, sorted, reclaimed, further processed, and converted into new recycled products. The analysis includes a variety of bales produced from sorting processes and the types of further processing that can prepare materials for downstream use. Among the processes studied are dissolution/precipitation and pyrolysis. Resource use and releases to the environment are described, along with the different products that can be made from these processes of recycled plastics.

[1] Smith, R.L., Takkellapati, S., and Riegerix, R.C. (2022), "Recycling of Plastics in the United States: Plastics Material Flows and Polyethylene Terephthalate Recycling Processes," *ACS Sustain. Chem. Eng.,* 10, 2084-2096. https://pubs.acs.org/doi/10.1021/acssuschemeng.1c06845

The views expressed in this presentation are those of the authors and do not necessarily represent the views or policies of the U.S. Environmental Protection Agency.

354d Achieving Social and Economic Justice Via Deployment of Plastic Waste Upcycling Technologies

Aurora Del Carmen Munguia Lopez, Marco A. Sánchez-Castillo, Victor Zavala

Plastics are versatile materials that have many societal, economic, and environmental benefits; for instance, they help preserve food, insulate buildings to save energy, and make vehicles more fuel-efficient [1,2]. However, commonly-used plastics are not naturally degradable and thus must be dealt with via recycling, incineration, or perpetual storage (in landfills or in the environment) [3]. Nearly 72% of all the plastic waste produced in the world requires perpetual storage [4]. The accumulation of plastics in the environment is largely influenced by mismanagement practices and causes ecological, social, and economic problems [5,6]. In 2019, it was reported that 22% of all the plastic waste produced in the world was mismanaged. However, strong variations in mismanagement are observed in developed and developing countries; for instance 4% of the plastic waste in the United States is mismanaged, while 42% is mismanaged in Latin America [4]. These discrepancies are due to differences in infrastructure, governmental policies, and education [7]. In developing countries, the lack of proper infrastructure and legislation gives rise to social and economic justice issues (e.g., creation of "informal" waste management sectors that exploit human labor). For instance, these informal sectors require self-employed "waste-pickers" that collect, haul, sort, and sell recyclables recovered from household waste and disposal sites. Currently, the main plastic waste recovered by pickers is polyethylene terephthalate (PET), which is sold to large corporations (e.g., beverage companies) [8]. Despite the large demand for PET in the mechanical recycling industry, waste-pickers do not receive a fair price for the recovered materials and thus have limited income [9]. As a result, waste pickers usually work long hours to achieve a living wage [10].

Advances in waste management and recycling technologies can help achieve a more equitable welfare distribution among all stakeholders involved. For instance, recent studies have highlighted that the deployment of pyrolysis technologies and coordinated management systems can open new and profitable value chains [11-13]. In this work, we explore the economic and social benefits of integrating pyrolysis technologies into the waste management infrastructure of Mexico. Our analysis uses a computational framework that integrates techno-economic analysis, value chain analysis, and socio-economic measures. We use this framework to establish conditions under which it would make sense to deploy pyrolysis systems and how this could lead to more equitable allocation of welfare to all stakeholders involved (including waste pickers). Our analysis reveals that the deployment of pyrolysis systems can lead to more fair labor and welfare allocation. Moreover, we show how this infrastructure can bring benefits to chemical companies in the United States, which can process plastic-based pyrolysis oil produced in Mexico to manufacture virgin polymer resins and thus contribute towards a more circular plastics economy.

[1] OECD (2022), Global Plastics Outlook: Economic Drivers, Environmental Impacts and Policy Options, OECD Publishing, Paris. https://doi.org/10.1787/de747aef-en.

[2] Evode, N., Qamar, S.A., Bilal, M., Barceló, D. and Iqbal, H.M., 2021. Plastic waste and its management strategies for environmental sustainability. Case Studies in Chemical and Environmental Engineering, 4, p.100142.

[3] Geyer, R., Jambeck, J.R. and Law, K.L., 2017. Production, use, and fate of all plastics ever made. Science advances, 3(7), p.e1700782.

[4] OECD (2022), Global Plastics Outlook Database. https://doi.org/10.1787/c0821f81-en

[5] Lau, W.W., Shiran, Y., Bailey, R.M., Cook, E., Stuchtey, M.R., Koskella, J., Velis, C.A., Godfrey, L., Boucher, J., Murphy, M.B. and Thompson, R.C., 2020. Evaluating scenarios toward zero plastic pollution. Science, 369(6510), pp.1455-1461.

[6] Law, K.L., 2017. Plastics in the marine environment. Annual review of marine science, 9, pp.205-229.

[7] Joshi, C., Seay, J. and Banadda, N., 2019. A perspective on a locally managed decentralized circular economy for waste plastic in developing countries. Environmental Progress & Sustainable Energy, 38(1), pp.3-11.

[8] Li, H., Aguirre-Villegas, H.A., Allen, R.D., Bai, X., Benson, C.H., Beckham, G.T., Bradshaw, S.L., Brown, J.L., Brown, R.C., Cecon, V.S. and Curley, J.B., 2022. Expanding plastics recycling technologies: chemical aspects, technology status and challenges. Green Chemistry, 24, pp.8899-9002

[9] Navarrete-Hernández, P. and Navarrete-Hernández, N., 2018. Unleashing waste-pickers' potential: supporting recycling cooperatives in Santiago de Chile. World Development, 101, pp.293-310.

[10] Botello-Álvarez, J.E., Rivas-García, P., Fausto-Castro, L., Estrada-Baltazar, A. and Gomez-Gonzalez, R., 2018. Informal collection, recycling and export of valuable waste as transcendent factor in the municipal solid waste management: A Latin-American reality. Journal of cleaner production, 182, pp.485-495.

[11] Ma, J., Tominac, P.A., Aguirre-Villegas, H.A., Olafasakin, O.O., Wright, M.M., Benson, C.H., Huber, G.W. and Zavala, V.M., 2023. Economic evaluation of infrastructures for thermochemical upcycling of post-consumer plastic waste. Green Chemistry, 25(3), pp.1032-1044.

[12] Kulas, D.G., Zolghadr, A., Chaudhari, U.S. and Shonnard, D.R., 2023. Economic and environmental analysis of plastics pyrolysis after secondary sortation of mixed plastic waste. Journal of Cleaner Production, 384, p.135542.

[13] Lubongo, C., Congdon, T., McWhinnie, J. and Alexandridis, P., 2022. Economic feasibility of plastic waste conversion to fuel using pyrolysis. Sustainable Chemistry and Pharmacy, 27, p.100683.

354e Microwave Catalysis Technology Application in Plastic Upcycling

Jianli Hu, Yuxin Wang

The chemical upcycling of plastic represents an attractive approach of using plastic waste as feedstock in promoting the circular economy and decarbonizing the chemical industry. Currently, the majority of plastic ends up in landfills or the environment, harming the ecosystem and affecting the natural environment. Thermochemical pathways (e.g., pyrolysis, gasification) and high pressure hydrogenolysis enable depolymerization films to produce small molecules that could be used as fuel or integrated into chemical refineries. However, these thermochemical processes suffer from severe challenges which include inevitably higher energy input leading to higher greenhouse gas (GHG) emission (i.e., ~12 $kgCO_2$/1 kg H_2 production in gasification); poor product selectivity; low value of products, thus requiring significant upgrading for converting to valuable products.

In this work, we reported the application of microwave catalysis technology application in plastic upcycling. Microwave heating has many advantages over conventional heating, including non-contact, selective and rapid heating, and quick start-up and shutdown. This fundamental difference results in important advantages in using microwave to initiate the catalytic depolymerization of plastics, particularly, the microwave irradiation's selectivity, as the microwave almost exclusively interact with the absorbing catalyst particles whilst the plastic substrate remains cold; thus, the self-decomposition of plastic polymers and many side reactions that occur at high temperatures are effectively limited.

The microwave catalysis technology demonstrated several advantages in plastic upcycling: (1) lower energy consumption leading to lower GHG emission, < 300 °C under atmospheric pressure; (2) higher selectivity of target product; (3) higher value of products (e.g., BTX aromatics (benzene, toluene, and xylene), ethylene, carbon nanofibers, and H_2, etc.); (4) moreover, the product can be adjusted based market requirement. For example, a high BTX aromatics (benzene, toluene, and xylene) yield of 44.7 % from low-density polyethylene (LDPE) was achieved using ZSM-5 catalyst at a low temperature of 300 °C under atmospheric pressure. Also, as shown in Figure 1, the high-quality separated CNFs can be generated from LDPE by microwave catalysis, meanwhile, the high purity H_2 (>93%) generated.

This proof-of-concept heralds an exciting new era of applications for the responsible and sustainable recycling of waste plastic, in which plastic are efficiently converted back to high-value petrochemical products.

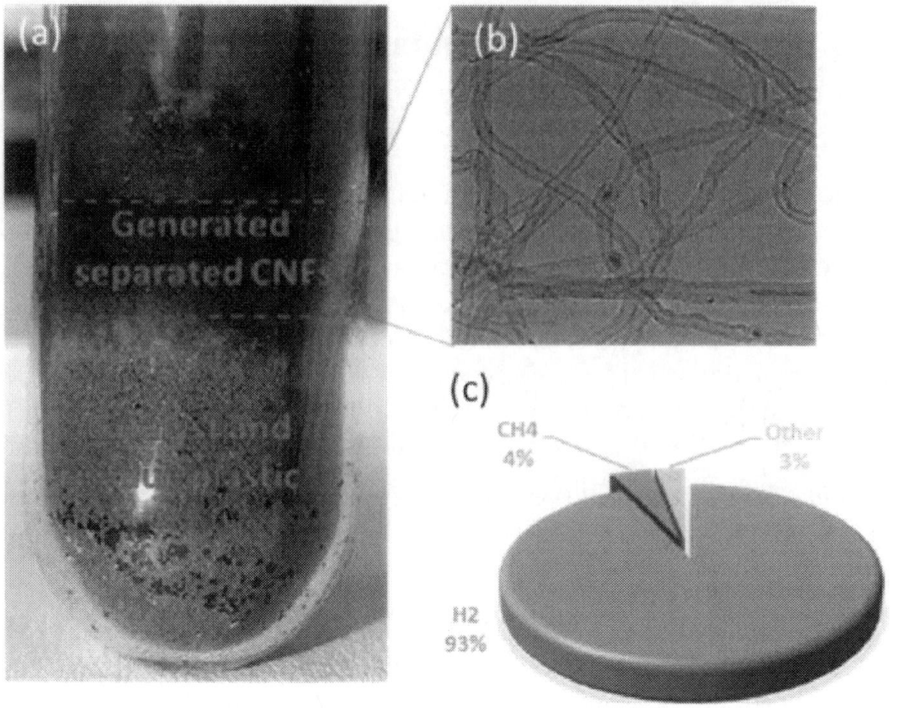

Figure 1, (a) Photo of after reaction, (b) TEM of produced CNFs, and (c) the composition of produced gas.

354f An Analysis of Plastic Additive Fates from the Generic Chemical Processing of End-of-Life Plastics

John D. Chea, Matthew Conway, Austin Lehr, Pahola Thathiana Benavides, Kirti Yenkie, Gerardo Ruiz-Mercado

Mechanical recycling has been widely used to recover end-of-life (EoL) plastics. However, the existing plastic recycling infrastructure requires further improvement to prevent unwanted releases of plastic additives from the polymer matrix, exposing humans and the environment to many hazardous substances. Chemical processing is a promising EoL treatment for handling discarded plastics because it can chemically deconstruct the macromolecules into new products and feedstock to synthesize the original material. This action can effectively separate chemical additives from the original matrix, provided that appropriate separation techniques are used. This work examines the fate of chemical additives throughout the conventional EoL chemical processing techniques, including pyrolysis, gasification, and depolymerization, by providing material flow, energy system, and generic scenario analyses. The material flow analysis shows that plastic additives could be released during polymer depolymerization and degradation because these molecules are not chemically linked to the original polymer chain. As a result, these chemicals may further react with substituents found inside the chemical recycling process units (e.g., reactors, dissolution tanks, etc.), synthesizing unwanted side products that may interfere with the quality of the final product and increase environmental releases of hazardous materials. The energy system analysis identified energy-saving potentials while ensuring the minimal generation of toxic chemical releases from each EoL chemical processing technique. Our generic scenario analysis estimates environmental releases and occupational exposures from operating a chemical recycling facility for plastics, along with the detailed EoL of all the materials involved, including monomers, additives, solvents, reagents, etc. In addition, this work evaluates the sustainability of the recycling process using The Gauging Reaction Effectiveness for the ENvironmental Sustainability of Chemistries with a multi-Objective Process Evaluator (GREENSCOPE). Therefore, this research demonstrates that chemical processing of the EoL plastics for potential reuse, although an excellent alternative to conventional mechanical recycling, requires further improvement to separate additives from plastics effectively and ensure the safety and sustainability of such technologies.

354g Techno-Economic Analysis of Non-Thermal Plasma-Assisted of Single-Use Plastics Film Wastes to Polyhydroxyalkanoates

Samirah Gnangbe, Harish Radhakrishnan, Xianglan Bai, Mark Mba Wright

The demand for plastics increases every year because of its wide application. Worldwide, plastics production reached over 390 million metric tons in 2021 and is estimated to increase by more than 50 percent in 2050 [1,2]. However, less than 9% of plastics produced get recycled or incinerated, and the remaining waste is dumped into landfills [3]. Single-use plastics film (SUPF), mainly composed of polyethylene (PE), represent the highest percentage of plastic waste sent to landfill because of the lack of technical and economical recycling processes [4]. Poor plastic waste management causes serious environmental problems, such as air, soil, and marine pollution.

Recent studies have investigated biopolymers, such as Polyhydroxyalkanoates (PHA) or Polylactic acid (PLA), as possible alternatives to mitigate the use of petroleum-based plastics because of their sustainability, ease biodegradability, compatibility, and climate-friendly production [5]. This study evaluates the techno-economic and feasibility of a non-thermal CO_2 plasma to upcycle mixed SUPF wastes into PHAs via pyrolysis oil fermentation.

A process model of the plasma degradation of SUPF waste was developed using *BioSTEAM.* A simulation was performed to estimate the plant's mass and energy balance, and it was assumed that 200 MT/day of plastics is processed in a day. The process takes place in four steps: Plasma reaction, PHA fermentation, PHA recovery, and oil recovery. The process outputs includes PHA and pyrolysis oil-derived chemicals. The mass balance showed conversion yields of 11.52 MT PHA, 51.51 MT alcohols, 16.1 MT paraffin, 11.56 MT olefins, 12.21 MT carboxylic acid, 7.61 MT day carbonyls, and 28.71 MT heavy oil (C29-44) per day, from 200 MT of plastics. For the economic analysis, we assumed three scenarios for the pyrolysis oil product value: 1) No-value, 2) Energy content-value, and 3) Market-value.

We estimated capital costs and the PHA minimum selling-price (MSP) based on three scenarios. The fixed capital investment is estimated at about $64 million. The PHA MSP ranges from $0.82 to $3.00/kg for the three scenarios. Sensitivity analysis indicates that PHA fermentation yield, SUPF plasma conversion, and plant size are the most important cost factors. Power consumption and electricity price have a small impact on the MSP. These findings suggest that plasma pyrolysis is a viable alternative for the upcycling of SUPF. Future work will investigate the lifecycle impacts of the technology.

References:

[1] Statista, 'Global plastic production 1950-2020,' Statista, https://www.statista.com/statistics/282732/global-production-of-plastics-since-1950/, Accessed on: Mar. 29, 2023.

[2] Statista, 'Global thermoplastic production forecast 2025-2050,' Statista, https://www.statista.com/statistics/1220981/thermoplastic-production-worldwide-forecast/, Accessed on: Mar. 29, 2023.

[3] H. Ritchie and M. Roser, 'Plastic Pollution,' OurWorldInData.org, 2018. [Online]. Available: https://ourworldindata.org/plastic-pollution.

[4] Y. Chen, A. K. Awasthi, F. Wei, Q. Tan, and J. Li, "Single-use plastics: Production, usage, disposal and adverse impacts," Sci. To*tal Environ.*, vol. 752, p. 141772, Jan. 2021, doi:10.1016/j.scitotenv.2020.141772.

[5] R. Sirohi, J. Prakash Pandey, V. Kumar Gaur, E. Gnansounou, and R. Sindhu, 'Critical overview of biomass feedstocks as sustainable substrates for the production of polyhydroxybutyrate (PHB),' Bioresour Technol, vol. 311, p. 123536, Sep. 2020. doi: 10.1016/j.biortech.2020.123536.

354h Environmental and Economic Analyses of Liquid Fed Pyrolysis Process for Waste Polyethylene Films

Utkarsh Chaudhari, Daniel Kulas, Alejandra Peralta, Tasmin Hossain, Anne Johnson, Damon Hartley, Robert Handler, Libby Umlor, Alayna Cronan, Barbara Reck, Vicki Thompson, David W. Watkins, David Shonnard

About 71% of total plastics production in the U.S. are polyethylene terephthalate (PET, resin code #1) and polyolefin plastics such as high-density polyethylene (HDPE, resin code #2), linear low-density polyethylene (LLDPE, resin code #4), low-density polyethylene (LDPE, resin code #4), and polypropylene (PP, resin code #5). These types of plastics also represent 79% of the total plastic waste generation in the U.S., with 77% being landfilled, 16% incinerated for energy recovery, and only 7% collected for recycling[1].

To break this linearity and dependence on limited fossil resources of plastic supply chains, emerging chemical recycling technologies have been proposed to replace fossil derived plastic resins. These technologies can be broadly classified in 3 main categories: dissolution/purification, depolymerization, and conversion recycling technologies[2]. Conversion recycling technologies convert plastic waste into intermediate hydrocarbon products such as specialty chemicals, waxes, lubricants, fuels, and gases. Pyrolysis, gasification, hydrothermal liquefaction, hydrogenolysis are some of the examples of thermal conversion recycling technologies. In this work, we look the environmental and economics analyses of a novel liquid-fed pyrolysis process that was developed at the Michigan Technological University[3,4]. We look at three different scenarios for technology, first, co-locating this process with a petrochemical facility ("integrated facility" scenario), and second, locating this process remotely and away from a petrochemical facility ("non-integrated facility" scenario), and third, increasing the yield of wax at a non-integrated facility. Based on our previously published systems analysis framework[2, 5], we take a "bottom-up" approach, in which process simulation is integrated with life cycle assessment (LCA) and techno-economic analysis (TEA). The **first research objective** of our study was to conduct process simulation of liquid fed pyrolysis processes. The **second research objective** was to evaluate technoeconomic performance metrics of the simulated recycling processes and compare these different scenarios mentioned above. We also look at the effect of scaling up the capacity on the total capital costs and minimum selling price of liquid naphtha, the main pyrolysis product. We also couple the economy of scale analysis with a simple transportation model to account for transportation costs over longer distances with increasing capacity, which has been neglected in the prior TEA studies for chemical recycling technologies. The **third research objective** was to evaluate environmental metrics such as GHG emissions and cumulative energy demand (CED) of simulated recycling processes and compare it against their fossil counterparts.

Our results revealed that the economic feasibility of this process depends on the capacity, location, product distribution, and price of pyrolysis products. For the integrated facility scenario, sales from pyrolysis gas product to the petrochemical facility favors the economics, whereas that for the non-integrated facility with high yield of wax, favors the economics. The MSP of pyrolysis oil produced at an integrated and non-integrated facility ranged from \$1.019 to \$1.918/kg of pyrolysis oil and was the highest for the remote facility. Preliminary LCA results revealed that the GHG emissions and CED impacts for the processes compare well with their fossil counterparts. The "cradle-to-gate" GHG emissions of pyrolysis oil ranged from 0.56-0.68 kg CO_2-eq/kg of pyrolysis oil, whereas those for pyrolysis gas ranged from 0.57-0.7 kg CO_2-eq/kg of pyrolysis gas. Similarly, the energy demand for pyrolysis oil ranged from 8.2-10.58 MJ/kg of pyrolysis oil, whereas that for pyrolysis gas ranged from

8.45-10.91 MJ/kg of pyrolysis gas.

References

1. Chaudhari, U.S., Johnson, A.T., Reck, B.K., Handler, R.M., Thompson, V.S., Hartley, D.S., Young, W., Watkins, D. and Shonnard, D., 2022. Material Flow Analysis and Life Cycle Assessment of Polyethylene Terephthalate and Polyolefin Plastics Supply Chains in the United States. *ACS Sustainable Chemistry & Engineering*, *10*(39), pp.13145-13155.

2. Chaudhari, U.S., Lin, Y., Thompson, V.S., Handler, R.M., Pearce, J.M., Caneba, G., Muhuri, P., Watkins, D. and Shonnard, D.R., 2021. Systems analysis approach to polyethylene terephthalate and olefin plastics supply chains in the circular economy: A review of data sets and models. *ACS Sustainable Chemistry & Engineering*, *9*(22), pp.7403-7421.

3. Kulas, D.G., Zolghadr, A., Chaudhari, U.S. and Shonnard, D.R., 2023. Economic and environmental analysis of plastics pyrolysis after secondary sortation of mixed plastic waste. *Journal of Cleaner Production*, *384*, p.135542.

4. Zolghadr, A., Kulas, D. and Shonnard, D., 2022. Evaluation of Pyrolysis Wax as a Solvent in Polyolefin Pyrolysis Processing. *Industrial & Engineering Chemistry Research*, *61*(30), pp.11080-11088.

5. Shonnard, D., Tipaldo, E., Thompson, V., Pearce, J., Caneba, G. and Handler, R., 2019. Systems analysis for PET and olefin polymers in a circular economy. *Procedia CIRP*, *80*, pp.602-606.

354i Plant-Wide Modelling of Microwave-Assisted Plastics Upcycling Process

Harish Damahe, Vishal Tuli, Md Emdadul Haque, Yuxin Wang, Jianli Hu, Debangsu Bhattacharyya

Large amount of plastics are getting accumulated in the waste handling facilities[1]. Among all types of plastics that are widely used, the low-density polyethylene is one of the difficult ones to convert to value added products or to recycle. Some of the leading methods that are being investigated for handling low-density polyethylene are thermal and catalytic pyrolysis, and gasification among others. However, these processes are energy-intensive and produces large number of products that are difficult to separate without incurring excessive cost. In additions, much of the products are low-value products with small yield of ethylene monomer. To address these issues, microwave (MW)-based technologies are being developed for conversion of low-density polyethylene to ethylene monomers and other monomers that can replace and supplement a large portion of the monomers being produced today from petroleum-derived feedstocks. Our in-house MW-assisted reactor yields high energy efficiency, achieved due to the low reaction severity, high conversion, achieved due to thermal and non-thermal effects of the MW reactors leading to high carbon recovery and utilization, and yield of value-added products, achieved by highly selective MW-enhanced catalytic depolymerization and aromatization process.

In this work, a kinetic model of the microwave-assisted, catalytic degradation of low-density polyethylene is developed[2]. The kinetic model yields products that are observed in an inhouse batch reactor. The reactor is ramped to the desired temperature following a temperature profile. Time-varying product composition from the lab reactor is analyzed. The laboratory data can fail to satisfy the mass balances and in particular for this process, can have error for carbon and hydrogen balances. If the raw data from the reactor are sued without satisfying mass balances, it can lead to incorrect estimate of the rate parameters in the reaction kinetic model. Therefore, dynamic data reconciliation is undertaken for exactly satisfying overall mass and atom balances in the experimental data. A kinetic mechanism is proposed based on the observation of product formation profile from the laboratory reactor. The reconciled dynamic data are utilized for optimal estimation of rate parameters by solving a nonlinear programming problem. The reactor model is scaled up to the commercial scale. A plant-wide model is developed for separation producing monomer-grade ethylene. The separation process has considerable differences in the operating conditions and sequence of distillation towers compared to the conventional ethylene production process using ethane crackers. Furthermore, no quench is necessary like the conventional process due to low temperature of the MW-assisted reactor. A cascaded refrigeration system is modeled for the separations section, first cooling with propylene refrigerant followed by ethylene refrigerant is considered. Following a sequence of towers, 99.2% pure ethylene that satisfies the monomer grade purity is obtained from the top of C2 splitter. Overall mass yield of ethylene per unit mass of recycled plastic is found to be more than 46%. A number of sensitivity studies are done by considering the time-varying product yield from the MW reactor and the operating conditions and sequence of the distillation towers.

References

1. https://www.epa.gov/facts-and-figures-about-materials-waste-and-recycling/guide-facts-and-figures-report-about
2. Chandrasekaran, B.Kunwar, B. Moser, N.Rajagopalan, and B.Sharma, "Catalytic Thermal Cracking of Postconsumer Waste Plastics to Fuels. 1. Kinetics and Optimization", *ACS Energy Fuels* 2015,29,6068-6077, doi: 10.1021/acs.energyfuels.5b01083.

354j Multi-Scale Regionalized Assessment of the Absolute Environmental Potential of Circular Polymers in Europe

Cecilia Salah, Ioan-Robert Istrate, Eldbjørg B. Vea, Anders Bjørn, Gonzalo Guillén-Gosálbez

The global demand for polymers has been increasing, notably in the plastic packaging sector, representing 31% of the annual plastic consumption in 2019 [1]. The sector heavily depends on fossil feedstock, with most polymers deriving from refinery operations, particularly the business-as-usual (BAU) technology, naphtha steam cracking [2]. Moreover, this industry follows a mostly linear economy model, with only a small fraction of the total plastic waste generated being recycled (19.5% in 2015) [3]. For instance, 95% of the value of plastic packaging is lost after their first use. They have a particularly short life span and generate substantial amounts of waste globally, i.e., 140 million tons (46.7% of the global plastic waste generation) in 2015 [2,3]. The need for a circular economy of polymers has motivated the development and implementation of new recycling technologies to extend the lifetime of materials and decrease the detrimental environmental effects caused by mismanaged plastic waste [4]. Schwarz et al. provided insights into the technology readiness level (TRL) of 12 recycling pathways and their respective carbon footprint [5]. Their results showcase the potential of deploying chemical recycling technologies (i.e., pyrolysis and gasification) for monomer production. They present lower emissions than waste-to-energy alternatives and can compete with mechanical recycling for most polymer types.

Although recycling technologies exist, waste collection and sorting are still challenging steps in the implementation of a circular economy of polymers. Sorting efficiencies and recovery rates depend heavily on polymer type, and pure streams of waste polymers can only be obtained after significant sorting efforts [6]. Additionally, despite the potential of deploying high TRL technologies to produce high-quality secondary materials, only a small fraction of the total European plastic consumption is covered by secondary plastics (5.5%), as shown by Hsu et al. [7]. This indicates that, despite the increasing recycling rates reported by European countries [8], the local polymer economy is still very linear.

Our study aims to quantify, for the first time, the current environmental impact of the polymer economy in European countries by combining process simulation with a regionalized life cycle assessment (LCA) analysis, thereby covering the multi-scale nature of the problem. The study is based on material flows of four main polymers used in plastic packaging, i.e., polyethylene (PE), polypropylene (PP), polystyrene (PS), and polyethylene terephthalate (PET). The assessment considers the production, consumption and waste management of polymers in EU countries. Furthermore, we compare the country-wise BAU scenarios to an alternative chemical recycling route, which produces secondary polymers from waste in a four-step pathway with three intermediate products (i.e., syngas, methanol, and olefins).

The chemical recycling pathway consists of a methanol-from-waste polymer plant coupled with a methanol-to-olefins unit, which in turn provides olefins as raw materials for the polymerization step to generate secondary polymers that can be re-implemented in the economy. The methanol-from-waste polymer process is simulated in Aspen Plus v12, while the two adjacent units are taken from literature, as they are well implemented in the chemical industry. The simulated methanol production occurs in two steps: first, a mixed-polymers stream reacts with steam in a gasifier to produce syngas, which then converts to methanol in a catalytic reactor. Carbon capture and storage (CCS) is deployed to minimize the process' CO_2 emissions. Data from the mass and energy balances are used to build the life cycle inventory of methanol-from-polymers.

To quantify the environmental impacts of the country-specific BAU system and the alternative circular model deploying chemical recycling, we perform a LCA for regional systems. This LCA bridges the process and technosphere levels by connecting the foreground data taken from the process model with the background data of the surrounding processes in different EU locations. The assessment is implemented in Brightway2 v2.4.2 using data from Ecoinvent 3.8, following a cradle-to-grave approach. The LCA method employed is based on the absolute environmental sustainability assessment (AESA) framework [9], which relies on the planetary boundaries (PBs) framework developed by Rockström et al. [10] and Steffen et al. [11]. This approach considers sustainability thresholds for different Earth-system processes beyond which technologies can be deemed environmentally unsustainable. For regional systems, the environmental impacts should be calculated according to the resource availability of the specific locations and the PBs adapted to local scales [12]. Thus, we here employ characterization factors described by Bjørn et al. [13], Pierrat et al. [14], and Vea et al. [15] to quantify the impacts of the studied systems on their respective locations. Moreover, we apply different downscaling methods to assess the environmental sustainability levels of the studied systems against regional boundaries defined for freshwater use, biogeochemical flows of nitrogen, and climate change.

Along these lines, our work combines process simulation with regional AESA to provide an overview of the current environmental performance of the BAU scenario of the country-specific polymer economy and the proposed circular solution. Preliminary results showcase that the BAU systems of the European countries transgress at least one of the regional boundaries and, therefore, can be deemed environmentally unsustainable. Moreover, the implementation of the proposed circular alternative can ease reducing the pressure on the studied Earth-system processes and contribute to a greener polymer economy in Europe. On the same note, the world demand for plastics will likely increase over time, accentuating the need for effective waste treatment methods. The proposed process serves as a possible recycling path to polymers that can be incorporated to satisfy both demands simultaneously.

Acknowledgments: This research is a part of the National Center of Competence and Research NCCR Catalysis, funded by the Swiss National Science Foundation.

References

1. **Statista**, Distribution of plastic consumption worldwide in 2019, by application, (2023), https://www.statista.com/topics/10136/plastic-packaging-industry-worldwide/ *[last accessed: 02.04.2023]*
2. **Ellen MacArthur Foundation**, The New Plastics Economy - Rethinking the future of Plastics, (2016)
3. **Geyer *et al.***, Production, use, and fate of all plastics ever made, *Science Advances*, 3(7), (2017)
4. **Ellen MacArthur Foundation**, Towards the circular economy, (2013)
5. **Schwarz *et al.***, Plastic recycling in a circular economy; determining environmental performance through an LCA matrix model approach, *Waste Management*, 121, 331-342, (2021)
6. **Kleinhans *et al.***, Development and application of a predictive modelling approach for household packaging waste flows in sorting facilities, *Waste Management*, 120, 290-302, (2021)
7. **T. Hsu *et al.***, How circular are plastics in the EU?: MFA of plastics in the EU and pathways to circularity, *Cleaner Environmental Systems*, 2, (2021)
8. **Plastics Europe**, Plastics – The Facts 2022, (2022)
9. **Bjørn *et al.***, A Framework for Development and Communication of Absolute Environmental

Sustainability Assessment Method, *Journal of Industrial Ecology*, 23, 4, (2019)

10. **Rockström *et al.***, Planetary Boundaries: Exploring the Safe Operating Space for Humanity, *Ecology and Society*, 14, 2, (2009)
11. **Steffen *et al.***, Planetary boundaries: Guiding human development on a changing planet, *Science*, 347, 6223, (2015)
12. **Ryberg *et al.***, Downscaling the planetary boundaries in absolute environmental sustainability assessments – A review, *Journal of Cleaner Production*, 276, (2020)
13. **Bjørn *et al.***, A planetary boundary-based method for freshwater use in life cycle assessment: Development and application to a tomato production case study, *Ecological Indicators*, 110, (2020)
14. **Pierrat *et al.***, Multicompartment Depletion Factors for Water Consumption on a Global Scale, *Environmental Science & Technology*, 57, 10, 4318–4331, (2023)
15. **B. Vea *et al.***, Spatially differentiated marine eutrophication method for absolute environmental sustainability assessments, *Science of the Total Environment*, 843, (2022)

AUTHOR INDEX

AIChE
120 Wall Street, FL 23
New York, NY 10005-4020

ISBN 978-1-7138-9312-7

CURRAN ASSOCIATES INC.
proceedings
.com